PASSIVE MICROWAVE REMOTE SENSING OF OCEANS

WILEY-PRAXIS SERIES IN REMOTE SENSING
Series Editor: David Sloggett, M.Sc., Ph.D.
Co-Director, Dundee Centre for Coastal Zone Research, UK
Deputy Chairman, Anite Systems, UK

This series aims to bring together some of the world's leading researchers working in the forefront of the analysis and application of remotely sensed data and the infrastructure required to utilize the data on an operational basis. A key theme of the series is monitoring the environment and the development of sustainable practices for its exploitation.

The series makes an important contribution to existing literature encompassing areas such as: theoretical research; data analysis; the infrastructure required to exploit the data; and the application of data derived from satellites, aircraft and *in situ* observations. The series specifically emphasizes research into the interaction of elements of the global ecosystem publishing high-quality material at the forefront of existing knowledge. It also provides unique insights into examples where remotely sensed data is combined with Geographic Information Systems and high-fidelity models of the physical, chemical and biological processes at the heart of our environment to provide operational applications of the remotely sensed data.

Aimed at a wide readership, the books will appeal to professional researchers working in the field of remote sensing, potential users of the information and data derived from the application of remote sensing techniques, and postgraduate and undergraduate students working in the field.

For further details of the books listed below and ordering information, why not visit the Praxis Web Site at http://www.praxis-publishing.co.uk

HANDBOOK OF SEAFLOOR SONAR IMAGERY
Philippe Blondel and Bramley J. Murton, Southampton Oceanography Centre, Southampton, UK

EARTHWATCH: The Climate from Space
John E. Harries, Professor of Earth Observation, Imperial College, London, UK

GLOBAL CHANGE AND REMOTE SENSING
Kirill Ya. Kondratyev, Research Centre for Ecological Safety, Russian Academy of Sciences, St Petersburg, Russia; A. A. Buznikov, Electrotechnical University of St Petersburg, Russia; O. M. Pokrovsky, Main Geophysical Observatory, St Petersburg, Russia

HIGH LATITUDE CLIMATE AND REMOTE SENSING
Kirill Ya. Kondratyev, Research Centre for Ecological Safety, Russian Academy of Sciences, St Petersburg, Russia; O.M. Johannessen, Nansen Environment and Remote Sensing Centre, Bergen, Norway; V.V. Melentyev, Nansen International Environmental Remote Sensing Centre, St Petersburg, Russia

MULTIDIMENSIONAL GLOBAL CHANGE
Kirill Ya. Kondratyev, Research Centre for Ecological Safety, Russian Academy of Sciences, St Petersburg, Russia

REMOTE SENSING AND GEOGRAPHIC INFORMATION SYSTEMS: Geological Mapping, Mineral Exploration and Mining
Christopher A. Legg, United Kingdom Overseas Development Administration, Forest and Land Use Mapping Project, Forest Department, Colombo, Sri Lanka

PASSIVE MICROWAVE REMOTE SENSING OF OCEANS
Igor V. Cherny, Center for Program Studies, Russian Academy of Sciences, Moscow, Russia, and Victor Yu Raizer, Space Research Institute, Russian Academy of Sciences, Moscow, Russia

SATELLITE OCEANOGRAPHY: An Introduction for Oceanographers and Remote-sensing Scientists
Ian S. Robinson, Department of Oceanography, University of Southampton, UK

REMOTE SENSING OF TROPICAL REGIONS
Eugene A. Sharkov, Remote Sensing Laboratory, Space Research Institute, Russian Academy of Sciences, Moscow, Russia

GEOGRAPHIC INFORMATION FROM SPACE: Processes and Applications of Geocoded Satellite Images
Jonathan Williams, Consultant, Space Division, Logica plc, Leatherhead, UK

Forthcoming titles

LIMNOLOGY AND REMOTE SENSING: A Contemporary Approach
Kirill Ya. Kondratyev, Research Centre for Ecological Safety, Russian Academy of Sciences, St Petersburg, Russia, and others

PASSIVE MICROWAVE REMOTE SENSING OF OCEANS

Igor V. Cherny
Head of Laboratory, Center for Program Studies,
Russian Academy of Sciences, Moscow, Russia

Victor Yu Raizer
Senior Scientist, Space Research Institute,
Russian Academy of Sciences, Moscow, Russia

JOHN WILEY & SONS
Chichester • New York • Weinheim • Brisbane • Singapore • Toronto

Published in association with
PRAXIS PUBLISHING
Chichester

Copyright © 1998 Praxis Publishing Ltd
The White House,
Eastergate, Chichester,
West Sussex, PO20 6UR, England

Published in 1998 by
John Wiley & Sons Ltd
in association with Praxis Publishing Ltd

Wiley Editorial Offices

John Wiley & Sons Ltd, Baffins Lane,
Chichester, West Sussex, PO19 1UD, England

John Wiley & Sons, Inc., 605 Third Avenue,
New York, NY 10158-0012, USA

Wiley-VCH Verlag GmbH, Pappelallee 3,
D-69469 Weinheim, Germany

Jacaranda Wiley Ltd, G.P.O. 33 Park Road, Milton,
Queensland 4001, Australia

John Wiley & Sons (Asia) Pte Ltd, 2 Clementi Loop #02-01,
Jin Xing Distripark, Singapore 12981

John Wiley & Sons (Canada) Ltd, 22 Worcester Road,
Rexdale, Ontario, M9W 1L1, Canada

Library of Congress Cataloguing-in-Publication Data

Raizer, Victor Yu.
   Passive microwave remote sensing of oceans / Victor Yu. Raizer, Igor V. Cherny
    p. cm.
   Includes bibliographical references (p. —) and index.
   ISBN 0-471-97170-7
   1. Oceanography—Remote sensing. 2. Microwave remote sensing.
   I. Cherny, Igor V. II. Title.
GC10.4.R4R39 1997
551.46'0028—dc21   97-28549
                                                 CIP

A catalogue record for this book is available from the British Library

ISBN 0-471-97170-7

Printed and Bound in Great Britain by MPG Books Ltd, Bodmin
Typeset by Heather FitzGibbon, 11 Castle Street, Christchurch, Dorset, BH23 1DP

Table of contents

The colour illustrations appear between pages 152 and 153.

Preface

This book reflects the most important stages of Soviet/Russian experimental and theoretical investigations in the field of ocean remote sensing, in particular with the use of radiometric microwave techniques and multifrequency aerospace instruments. Great experience and large bodies of field data were collected in the period 1974–1992 at the Space Research Institute, Moscow, where the authors have worked over many years under the leadership of Professor V.S. Etkin (1931–1995). These investigations included the development of the theoretical models, creation of experimental methods and instruments, and data processing software. As a result of these investigations, the scientific lead, or 'Radiohydrophysics', was established. It means that the diagnostics of ocean–atmosphere dynamics and hydrodynamic processes on the ocean surface and in deep waters can be reliably evolved by using compact passive radiometric sensors. Suffice it to say that the energetic capacity of the modern aerospace multifrequency radiometer is lower than that of a similarly active single-frequency radio system or radar.

However, the physical background of ocean microwave radiometry is more complex than the well-developed theories of scattering and radar glitter from the oceans, since in our case both electromagnetic propagation theory and the thermohydrodynamic nature of the ocean–atmosphere interface must be considered simultaneously. In fact, we often have to deal with the slowly contrasting spatial–temporal variances of the environment's own emission that reflect, in particular the ocean–atmosphere conditions, as a single geophysical object. Therefore, the main problem of ocean microwave diagnostics is the creation of a theoretically justified concept, needed to advance aerospace microwave technology and data processing tools.

At the present time, there is no known universal theory or computer methods for the solutions of airspace problems, associated with ocean remote sensing. Complex influence of such natural factors as multiscale surface waves and roughness disturbances, two-phase bubble–foam–spray structures, thermohydrodynamic boundary effects, wind field, and surface temperature fluctuations cannot be taken into account in a unified theory or model. These factors may also associate with ocean–atmosphere mass and heat flux exchange and deep-ocean processes. However, it is possible to consider the influence of each oceanic factor on the ocean microwave emission individually. The book is a partial guide to these questions.

The main material of the book was published in the Russian edition with the title *Microwave Diagnostics of Ocean Surface*, by V.Yu. Raizer and I.V. Cherny (Gidrometeoizdat, Saint-Petersburg, 1994, 231 pp.). At the same time, the suggested English version of the book is not a direct translation of the Russian original. The text was essentially restructured and supplemented thoroughly in the sections concerned with the descriptions of modern microwave technology, multifrequency radiometric imagery, and possible applications of microwave techniques for deep-ocean remote sensing applications. At the same time, the sections concerned with the theory of electromagnetic wave propagation in random media such as the ocean boundary layer, and hydrodynamics of wave motions, non-linear wave–wave interactions, and wave-breaking processes were changed. In addition, a more adequate theoretical–experimental analysis of the ocean microwave effects was made.

The authors participated over a long period in a number of ship and aircraft experiments that were made with the fine collaboration of Dr V.P. Shevtsov, Pacific Oceanographic Institute, Russian Academy of Sciences, and Prof. G.Ya. Gus'kov, Scientific & Production Corporation "ELAS".

Now, the authors plan to realize practically many said scientific aspects and methods in the Microwave Oceanographic Satellite Program 'Meteor-3M' of the Russian Space Agency under the leadership of Prof. G.M. Chernyavsky, Director of the Center for Program Studies.

The authors wish to express their heartfelt thanks to a number of colleagues whose work, interest and attention have stimulated the investigations and the performance of field experiments and measures, in particular Prof. K.D. Sabinin, Prof. S.S. Moiseev, Prof. E.A. Sharkov, Prof. Yu.A Kravtsov, Dr S.S. Semenov, and Dr A.S. Petrosyan.

The authors would like to express their gratitude to officials of NASA Marshall Space Flight Center, Global Hydrology and Climate Center, and personally to Mr Michael Goodman, manager of the Global Hydrology and Resource Center, for suggestions of SSM/I DMSP data. Some of the SSM/I images have been included in this monograph.

The authors would like to express their gratitude to Clive Horwood, chairman of PRAXIS Publishing Ltd., who has expressed great interest in publishing an edition of the book in English; also, to all European and U.S. Reviewers for their useful comments and recommendations. To many other Russians and American people and scientists who have contributed in one way or another, the authors offer their thanks.

Moscow and Washington, September 1997

1

Introduction

The majority of large-scale processes and phenomena occurring at the interface of the ocean and atmosphere cannot be studied by using traditional ship *in situ* measurements. Among these are: the generation of wind-waves and storms, non-linear wave–wave interactions, variation of frontal zones and currents, oil-slicks spreading, and some synoptic phenomena.

Modern methods of remote sensing enable the problem of global geophysical monitoring, including the ocean–atmosphere system, to be tackled. The arsenal of aerospace remote sensing techniques used covers a wide range of electromagnetic waves, from ultraviolet waves to radiowaves. Microwaves include millimeter, centimeter and decimeter radiowaves. Microwaves possess a higher sensitivity to the structural variation of the natural medium and are absorbed weakly in the atmosphere and clouds.

As to the physical principle, microwave remote sensing techniques are divided into two different types: active and passive. The first (radar) uses the measurements of the scattering signal from the investigated medium to obtain information about its (sub)surface condition. The second (radiometric) is based on the measurements of the thermal emission of the medium, which is dependent on its thermodynamical and physical properties.

Radar techniques are quite well developed and are applied to solve different kinds of geophysical problems, for instance, the data from the SEASAT, ERS-1, ALMAZ, RADARSAT satellites. Radiometric techniques, however, are implemented less frequently and are used generally for the determination of atmosphere and cloud parameters, for the estimation of ocean surface temperature and wind speed, and also to monitor sea ice and land; for example, the well-known satellite data from the U.S. Special Sensor Microwave Imager (SSM/I). The possibilities for application of microwave radiometry to study ocean dynamics and surface wave perturbations are limited because of low spatial resolution of microwave equipment and, more importantly, an incomplete understanding of the physical mechanisms of thermal emission formation and microwave propagation at the ocean–atmosphere interface, under the influence of different hydrodynamical processes.

For studies of global, large-scale and local processes in the ocean connected with non-linear wave–wave interactions and wind-wave dynamics and breaking waves, utilization of multi-frequency microwave measurements is insufficient. Only one method remains for the visual fixing of the spatial–statistical characteristics of the ocean surface: via high qualitative aerospace optical photography. Thus, the combination of microwave and optical techniques is the most informative way to study the spatial structure and dynamics of the ocean surface, embracing scale measurements from centimeters to several hundreds of meters.

Prior to the present investigation, there were only general insights into variances of the microwave emission of the ocean surface. Previously, possibilities of the one- and two-scale models to calculate scattering and emission of the sea surface were analyzed. Estimations of the foam effect at microwave frequencies were made by using the simplest model, such as an air–water mixture and a two-layer dielectric medium. The first radio-metric measurements of the dependence of the brightness temperature on the wind speed at several fixed wavelengths ($\lambda = 0.8$–8 cm) are well known from the literature.

Along with the tropics, an overall understanding of the problem of the effectiveness of microwave measurements of the ocean surface was missing. It was unknown as to what phenomena in the ocean could be observed using aerospace microwave techniques, and what could not be observed.

Problems of electrodynamics at the ocean–atmosphere interface in storm conditions, when the wave-breaking and foaming processes are going on all the time, were not investigated. Theoretical and experimental problems of the scattering and emission in a strong non-homogeneous multiphase bubble–foam–spray medium were also not studied. The influence of the geometry and statistics of surface waves and small-scale hydrody-namical disturbances on the variations of the microwave emission were not satisfactorily investigated. Also statistical properties of the wind-wave-breaking field and variations of the wave number spectrum of wind waves at different ocean surface conditions were not studied. At the same time, these phenomena are the major hydrodynamic factors which determine the features of the microwave emission and the scattering of the ocean surface. In addition the methodology of natural remote sensing measurements, methods of signal processing, and physical interpretation of the radiometric data were not developed.

The objective of this monograph is the creation of physical principles of aerospace microwave radiometric remote sensing, and monitoring of multi-scale hydrodynamic processes and small disturbances in the ocean, including surface roughness and wind waves, internal waves, and foam/whitecaps coverage. The development of a modern hydrodynamics–electromagnetic model should be based on the consideration of the following main processes:

- Structure and dynamics of the ocean–atmosphere interface like an electrodynamical non-homogeneous medium.
- Mechanisms and conditions of microwave propagation, scattering, and emission of the ocean–atmosphere interface caused by combined geometrical and volume non-uni-formity.
- Change of spatial–temporal multi-scale macrostructure of the ocean surface under the influence of non-linear dynamical processes including wave–wave interaction and wave breaking.
- Influence of oceanic and atmospheric internal waves.

An important part of the research is reported in Chapter 5, which describes the microwave features of different hydrodynamical non-uniformity and multi-scale disturbances, originating on the ocean surface and atmosphere boundary layer.

The development of microwave remote sensing techniques in the future will be associated with the space monitoring of different oceanic phenomena, including the observations of wind-generated waves, storm surges, manifestation of internal waves, surface currents, thermal fronts, and small-scale hydrodynamic disturbances. It seems likely that multi-frequency microwave passive imagery with improved (~ several kilometers from space) spatial resolution will give new useful information about ocean dynamics and will be a top-level tool for expeditious ocean–atmosphere diagnostics.

2

Ocean surface phenomena

2.1 STRUCTURE OF THE OCEAN–ATMOSPHERE INTERFACE

The ocean–atmosphere interface presents a non-uniform multi-scale structure with variable hydrodynamic parameters. The current status of research on the development of a radiohydrophysical model can be illustrated by Figure 2.1. The main factor is perturbations of the air–water interface that adjust the level of the passive/active microwave signal and corresponding spectral and polarization characteristics of ocean microwave emission and scattering. Although a great number of theoretical and experimental investigations in the field of electrodynamics of rough surfaces and non-uniform multi-phase media are made, there is no statistically reliable microwave model that can describe the real temporal–spatial dynamics of the natural ocean surface at different hydrometeorological conditions and spectral features of the microwave emission simultaneously. The solution may be based on detailed information about the contribution of the individual oceanic attribute or factor. The suggested classification in general reflects this view: the ocean–atmosphere interface is represented as a multi-factor statistical and dynamical system with spatially distributed parameters and relationships.

The elements and their relationships may be ambiguous and must be specified and adjusted as knowledge is acquired about hydrophysical phenomena occurring on the ocean surface. But using even such information, a simplified model requires the creation of combined multiparameter algorithms. Thus, the inverse problem of microwave diagnostics of the ocean is *a priori* not correct mathematically, and therefore its solution requires additional useful information about established processes or phenomena.

It is important to note that introducing different elements and relationships into the model can be done spontaneously according to a certain law by using a set of statistical data. For example, geometrical factors such as for multi-scale surface waves must always be considered, but volume factors such as for bubbles, foam, and spray need to be considered only at specific situations. Consequently at the creation of hydrodynamic–electromagnetic concept, we come to the typical problem of numerical modeling (simulation) with a large set of incoming hydrophysical parameters. Let us consider the main hydrophysical factors.

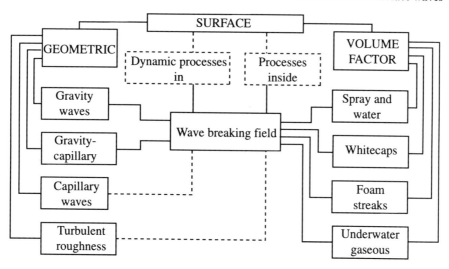

Figure 2.1. Main elements of the radiophysical model of the ocean surface.

2.2 CLASSIFICATION OF SURFACE WAVES

Ocean waves are strangely diversified by form and space–time scales. For this reason, two approaches are used to describe them. The first, deterministic, is based on the application of the strict hydrodynamic theory. It describes the profile of regular linear or non-linear waves on deep or shallow water. The second, statistical, operates with the probabilistic laws of distribution of energy between different wave components. In this case, it is presumed that the surface configuration fluctuates randomly in space and time.

More universal methods of describing surface waves which unite both approaches are connected with the numerical solution of equations of non-linear hydrodynamics and balance of the spectral density of energy. The most important result of the hydrodynamic theory is receiving numerical profiles of two-dimensional and even three-dimensional non-linear surface waves and the modeling of their evolution in space up to the breaking moment. It is also possible to investigate the criteria of instability, and the phenomenon of the bifurcation of gravity waves at their interaction.

In hydrodynamics, steady and non-steady-state surface waves are distinguished. Steady-state waves do not change their properties in space and time. Otherwise the waves are called non-steady-state waves. In addition, periodical linear and non-linear steady surface waves are separated (Table 2.1).

The most important type is the surface gravity waves of finite amplitude (Stokes waves). These waves are unsteady with respect to small periodic disturbances (Benjamin-Feir modulation instability). The effects of instability and evolution of one-dimensional and two-dimensional surface non-linear gravity waves in deep water have been investigated in detail [1, 2]. Recent progress in the non-linear theory of wave motions in fluid flows has been dramatic in the understanding of non-linear wave–wave interaction phenomena, in particular, the role of multi-mode wave–wave interactions, secondary

Table 2.1. The main types of steady-state surface waves (by theory)

Type of surface waves	Author
1. Linear periodical	Nekrassov, 1951
2. Trochoidal	Gerstner, 1802
3. Non-linear periodical with a fine amplitude	Stokes, 1847
4. Gravity solitary (soliton)	Boussinesq, 1890
5. Capillary linear periodical	Rayleigh, 1876; [Sekerzh-Zenkevich], 1972 [Emik, Toland], 1981
6. Capillary non-linear periodical	Crapper, 1957
7. Capillary solitary (soliton)	Monin, 1986

modulational instabilities and amplification mechanisms in the processes of generation of two- and three-dimensional 'chaotic' and/or 'coherent' hydrodynamic structures causing the fully developed oceanic turbulence [3, 4].

Flat weakly non-linear waves are described by the well-known Korteweg–de Vries (1895) equation. The solutions of this equation can be as periodical as solitary waves (solitons). The existence of gravity–capillary solitons in shallow water was proved theoretically on the basis of solutions of the non-linear Kadomtsev–Petviasvilli and Schroedinger equations.

Capillary waves or ripples are essentially non-linear. The theoretical profile of capillary waves has a complex and ambiguous form. Short capillary waves in the ocean are strong and unsteady. Although they are not waves, they can be represented by a random field of perturbations—like a field of short impulses or injections.

2.3 GENERATION AND STATISTICS OF WIND WAVES

Well-known mechanisms of generating surface roughness and formatting stationary wind-wave spectra in the ocean are:

- Due to surface wind stress.
- Kelvin–Helmholtz instability due to local wind shear.
- The Miles shear instability due to the influence of a matching layer with wind profile.
- Resonant mechanism due to non-linear interactions of gravity waves when the speeds of wave propagation and wind are the same [5].
- Weak turbulence theory [6] due to the locality of wave–wave interaction in the case of wind-driven wave turbulence.

In recent years the approach of slow dispersion and non-linearity of deterministic surface gravity–capillary waves has been developed. Using this theory, new solutions of the Korteweg–de Vries equation of the dynamics of solitons and their interactions were investigated [3].

Another mechanism of surface roughness generation deals with non-linear wave–wave interactions. The dynamics of this interaction are described by the kinematic theory for the statistical ensemble of surface waves [7]. This theory describes the formation of wavenumber spectrum in the oceans. An important application of the theory is the consideration of surface wave–current interactions. In particular, the effects of blocking gravity–capillary wave by surface currents induced by internal waves are manifested. As a result, the strong transformation of surface wavenumber spectra in the region of decimeter wavelengths has occurred. In the case of linear theory, the amplitude of surface waves decreases when they are propagated along the current, but the amplitude increases when waves are propagated against the current. It is possible to observe both an increase and a decrease in the wave energy's spectral density. An example is the propagation of surface waves on horizon-non-uniform currents in the field of oceanic internal waves. The theory explains the effects of anomalous roughness such as surface smoothing, slicks and 'rip currents'.

However, an exact universal all-purpose formula, which describes two-dimensional wavenumber spectra of ocean surface waves in the wide range of a wind wave's spatial frequency, does not exist. Therefore some empirical models of the spatial spectrum in different spectral intervals are used. For active/passive microwave remote sensing applications, the following presentation of fully developed wavenumber spectra is suggested.

According to the well-known empirical approximations and empirical relationships [8–13] the energetic part of full wavenumber spectra can be separated into the following five regions:

I. Region of large energy-carrier quasi-linear gravity waves (the Pierson & Moskovitz spectrum):

$$F_1(K) = 4.05 \cdot 10^{-3} K^{-3} \exp\left\{-0.74 g^2 / \left[V^4(u_*)K^2\right]\right\} \tag{2.1}$$

for the interval $0 < K < K_1 = K_2 u_{*m}^2 / u_*^2$, where

u is the wind speed at an altitude of 19.5 m (m/s);

$u_* = \sqrt{C_n V^2}$ is the friction velocity (cm/s);

$C_n = (9.4 \cdot 10^{-4} V + 1.09) \cdot 10^{-3}$ is the aerodynamic coefficient of drag;

$u_{*m} = 12$ cm/s.

II. Region of non-linear short gravity waves:

$$F_2(K) = 4.05 \cdot 10^{-3} K_1^{-1/2} K^{-5/2} \tag{2.2}$$

for the equilibrium interval $K_1 < K < K_2 \approx 0.359$ cm^{-1}.

III. Transfer region of dynamical equilibrium:

$$F_3(K) = 4.05 \cdot 10^{-3} D(u_*) K_3^{-P} K^{-3+P}, \tag{2.3}$$

$$\rho = \log\left[u_{*\mathrm{m}}D(u_*)/u_*\right]/\log(K_3/K_2),$$

$$K_2 < K < K_3 \approx 0.942 \text{ cm}^{-1},$$

where

$$D(u_*) = \left(1.247 + 0.0268u_* + 6.03\cdot10^{-5}u_*^2\right)^2 \qquad (2.4)$$

(the Pierson & Stacy approximation); or

$$D(u_*) = 1.0\cdot10^{-3}u_*^{9/4} \qquad (2.5)$$

(the Mitsuyasu & Honda approximation).
   Another form is:

$$F_3(K) = F_4(K_3)\left(K/K_3\right)^q, \qquad (2.6)$$

$$K_2 < K < K_3 \approx 0.942 \text{ cm}^{-1},$$

$$q = \log\left[F_2(K_2)/F_4(K_3)\right]/\log(K_2/K_2),$$

where $F_4(K_3)$ corresponds to (2.7).

IV.   The equilibrium range of the Phillips' spectrum of limiting gravity–capillary waves:

$$F_4(K) = 4.05\cdot10^{-3}D(u_*)K^{-3}, \qquad K_3 < K < K_\nu, \qquad (2.7)$$

$$K_\nu = 0.5756u_*^{1/2}\left[D(u_*)\right]^{-1/6}K_\mathrm{m},$$

$$K_\mathrm{m} = \left(\rho_\mathrm{w}g/\gamma_0\right)^{1/2} \cong 3.63 \text{ cm}^{-1},$$

where g is gravity, ρ_w is density of the water, and γ_0 is the surface tension coefficient.
   Another form is

$$F_4(K) = 0.875(2\pi)^{\rho_1-1}\,\frac{g+3g\,K^2/13.1769}{\left(gK+gK^3/13.1769\right)^{\frac{\rho_1+1}{2}}}, \qquad (2.8)$$

$$K_3 < K < K_4,$$

$$\rho_1 = 5.0 - \log u_*,$$

where K_4 is defined from the equation

$$F_4(K_4) = F_5(K_4). \qquad (2.9)$$

V.   Region of capillary waves and weak turbulence:

$$F_5(K) = 1.479\cdot10^{-4}u_*^3K_\mathrm{m}^6K^{-9}, \qquad (2.10)$$

$$K_\nu < K < \infty.$$

The results of calculations of the full wavenumber spectrum $F(K)$ using relations (2.1)–(2.10) are shown in Figure 2.2. The spectrum is parametrized by wind speed V.

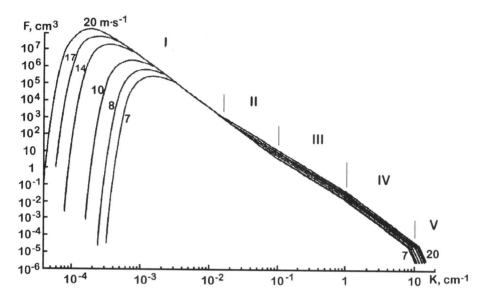

Figure 2.2. Numerical model of wavenumber spectra evolution. The value of wind speed changes from 7 to 20 m/s. Five spectral intervals are sewn together.

The statistical description is based on integral information about the change of averaging spectral density of wave energy, due only to slow variations in wind speed and interaction between the air flow and the ocean surface. Moreover, the wave field in the ocean is a multiple-scale non-linear dynamic system, which is characterized by a large degree of freedom. Resonant and non-resonant groups of waves exist in such a system. As it follows from a general theory, non-resonant wave–wave interactions provide a stable spatial evolution of the system. But resonant wave–wave interactions under certain conditions lead to different instabilities, which in time lead to chaos and noise.

The fundamental description of atmosphere and ocean dynamics is based on the Navier–Stokes equations:

$$\frac{\partial \mathbf{V}}{\partial t} + (\mathbf{V} \cdot \nabla)\mathbf{V} = -\frac{1}{\rho_t} \nabla P + v_0 \nabla^2 \mathbf{V} + \mathbf{F}, \tag{2.11}$$

$$\nabla \cdot \mathbf{V} = 0,$$

where $\mathbf{V}$ is the velocity vector and $\mathbf{P}$ is the pressure vector at each point $\mathbf{F}$ and instant t; ρ_t is the fluid density; v_0 is the kinematic viscosity; and $\mathbf{F}$ is the force term (gravity, stirring). Usually the solid boundaries or free surface and fluid boundaries are considered. Therefore, in the common case both non-linear kinematic and dynamical boundary conditions may be introduced together with equations (2.11).

Investigations of the non-linear equations (2.11) show that two principal types of solution can be found—stable and unstable. This means that in the classic understanding, 'real motions must not only satisfy the equations of hydrodynamics, but must be stable in the sense that the perturbations which inevitably arise under actual conditions must die out with time' [14]. It is clear that such a suggestion imposes stringent limits on the relation of an initial medium parameter and the non-linearity degree of the equations. Usually the Reynolds number $Re = U_0 L / \nu_0$ (where U_0 is the characteristic velocity, L is the characteristic scale, and ν_0 is the kinematic viscosity) is applied as the main criterion of the stability. The stable or unstable regimes of motion are determined by the value of the critical Reynolds number Re_c. If $Re < Re_c$, the regime is stable; if $Re > Re_c$, the regime may be unstable. Also, this criterion is used to estimate the ratio of the non-linear terms to the dissipative terms in the Navier–Stokes equation.

In the common case of wave–wave interactions, including wave–current interaction, the evolution of the spectral density of wave energy is described by the kinematic equation [7]:

$$\frac{\partial N}{\partial t} + (\mathbf{U} + \mathbf{C}_g)\nabla N = I_{in} + I_{nl} + I_{ds}, \tag{2.12}$$

where $N(\mathbf{K},\mathbf{r},t) = \rho \dfrac{\omega_0}{|\mathbf{K}|} S(\mathbf{K},\mathbf{r},t)$ is the action spectral density, $\mathbf{C}_g$ is the local group velocity, $\mathbf{U}$ is the current velocity vector, and $S(\mathbf{K},\mathbf{r},t)$ is the two-dimensional wave-number spectrum.

The processes which modify the action spectral density are described by the net source function $I_s = I_{in} + I_{nl} + I_{ds}$ on the right side of the equation. The source function is represented as the sum of the three terms: the energy flux due to non-linear resonant wave–wave interactions I_{nl}; and the energy loss due to wave breaking and other dissipative processes I_{ds}. In the case of the non-uniform surface current field, induced, for example, by internal wave packets, the dispersion relation for the surface waves may be written as

$$\omega(\mathbf{K},\mathbf{r},t) = \omega_0(\mathbf{K},\mathbf{r}) + \mathbf{K}\mathbf{U}, \tag{2.13}$$

where $\omega_0(\mathbf{K},\mathbf{r})$ is the dispersion relation for initial (non-disturbed) surface waves.

The spectral function of perturbation associated with any hydrodynamic process (for example, internal waves or surface currents) may be written as

$$f(\mathbf{K},\mathbf{r},t) = \frac{S_f(\mathbf{K},\mathbf{r},t) - S(\mathbf{K})}{S(\mathbf{K})}, \tag{2.14}$$

where $S(\mathbf{K})$ is the initial (non-disturbed) wavenumber spectrum.

On the basis of equations (2.11)–(2.14), in principle it is possible to calculate the perturbation spectrum $S_f(\mathbf{K},\mathbf{r},t)$ and use it as an input parameter in the ocean microwave models and applications. This approach was used for the quantitative analysis of radar signatures of surface waves and internal waves in the JOWIP and SARSEX field experiments (*J. Geophys. Res.*, 1988, **93**, C10 and C11). Obviously, a similar description may also apply for interpretation of radiometric microwave signatures, associated with the influence of surface roughness disturbances.

Analysis of the action balance equation has been carried out by many authors [5, 6, 15]. In particular, problems of surface modulation and surface wave–current interaction in the internal wavefield were investigated in detail. However, only simple approaches, when the source function equaled $I_s = 0$, $I_s = I_{in}$ or $I_s = I_{nl}$, were considered. Moreover, the universal character of the full dynamic non-linear equations (2.11) and (2.12) permit modeling any dynamic situation numerically, including the possible progress of instabilities, and generation of two-dimensional surface structures.

Non-linear, wave–wave interactions can be separated into two types: weak and strong. The first type of synchronous interactions is first-order non-linear effects for waves of finite amplitude and slope. Non-linearity causes a slow change in the wave characteristics in space and time, and provides small perturbations. This process is characterized by long duration of the interactions. The second type is characterized by small time and small spatial scales of interactions. In this case, different types of instability are advanced. The strong interactions cause, for example, the wave-breaking phenomena.

For second-order resonant interactions among a triad of surface waves, the following conditions of synchronism must be satisfied simultaneously,

$$\mathbf{K}_1 = \mathbf{K}_2 + \mathbf{K}_3, \quad \omega_1 = \omega_2 + \omega_3, \quad \omega = (gK)^{\frac{1}{2}}, \tag{2.15}$$

where K and ω are the wave number and frequency of wave components. There are no non-trivial solutions of the equation (2.15). But resonance cannot occur at this order, and only the effect of the perturbation of the wave profile can be seen [5].

The interaction of the three wave components $(\mathbf{K}_1, \mathbf{K}_2, \mathbf{K}_3)$ at the quadratic and cubic orders generates components with the numbers $(\mathbf{K}_1 \pm \mathbf{K}_2 \pm \mathbf{K}_3)$. For resonance among a tetrad of wave components, the conditions of synchronism must be satisfied or nearly satisfied,

$$\mathbf{K}_1 \pm \mathbf{K}_2 \pm \mathbf{K}_3 \pm \mathbf{K}_4 = 0, \quad \omega_1 \pm \omega_2 \pm \omega_3 \pm \omega_4 = 0, \quad \omega = (gK)^{\frac{1}{2}}. \tag{2.16}$$

The non-trivial solution of the equation (2.16) exists for four-wave interactions

$$\mathbf{K}_1 + \mathbf{K}_2 = \mathbf{K}_3 + \mathbf{K}_4, \quad \omega_1 + \omega_2 = \omega_3 + \omega_4, \quad \omega = (gK)^{\frac{1}{2}}. \tag{2.17}$$

This scheme describes the four-wave interactions of weakly non-linear surface gravity waves on a deep water [1]. This interaction mechanism causes energy transfer in the space–time spectrum, and affects its broadening at wind-wave generating conditions [7].

There is an important and particular case of a four-wave interaction model when two of the primary wave numbers are coincident $(\mathbf{K}_3 = \mathbf{K}_4)$. The resonant conditions (2.17) change as:

$$\mathbf{K}_1 + \mathbf{K}_2 = 2\mathbf{K}_3, \quad \omega_1 + \omega_2 = 2\omega_3. \tag{2.18}$$

These conditions were tested and investigated experimentally in a laboratory when the wavenumber vectors $\mathbf{K}_1$ and $\mathbf{K}_2$ were perpendicular [5]. But under open ocean conditions, strict satisfaction of the resonance for several systems of surface waves is impossible. The phenomena of quasi-synchronism due to non-stationary and non-coherent interaction between weakly non-linear gravity waves were investigated by using satellite, airborne radar and optical remote sensing data [16–18]. In the ocean it is possible to observe quasi-resonant wave components which satisfy the conditions,

$$\mathbf{K}_1 + \mathbf{K}_2 = \mathbf{K}_3 + \mathbf{K}_4 - \Delta\mathbf{K}, \text{ or}$$

$$\mathbf{K}_1 + \mathbf{K}_2 = 2\mathbf{K}_3 - \Delta\mathbf{K}, \quad \text{or} \quad 2\mathbf{K}_1 - \mathbf{K}_2 = \mathbf{K}_3 + \Delta\mathbf{K}, \tag{2.19}$$

where $\Delta\mathbf{K}$ is the phase mismatch. The value of the phase mismatch characterizes the group structure of interacting waves and depends on the extent of non-stationarity or non-uniformity of the wave-generating system investigated.

The weak turbulence theory is based on the solution of the kinetic equation for spatial wavenumber spectra or the special density of the wave action. This equation accounts for four-wave resonant interactions if the dispersion law is of the non-decay type, such as surface gravity waves, and three-wave interactions for the decay-type laws like capillary waves. If a wave field is statistically isotropic, these equations have exact stationary solutions in the form of power laws like the Kolmogorov spectra.

For surface gravity waves, two solutions exist in terms of the spectral density of wave energy. The first solution is

$$F(K) = \alpha_q g^{-\frac{1}{2}} q^{\frac{1}{3}} K^{-\frac{7}{2}}, \tag{2.20}$$

where q is the energy flux down the spectrum and α_q is the non-dimensional constant. The second solution is

$$F(K) = \alpha_q p^{\frac{1}{3}} K^{-\frac{10}{3}}, \tag{2.21}$$

where p is the action flux up the spectrum. These spectra were tested using oceanographic (*in situ*) measurements and remote sensing data.

In the case of capillary waves there are only wave–wave resonant triplets. The corresponding Kolmogorov spectrum is

$$F(K) = \frac{3}{2}\alpha_\sigma q^{\frac{1}{2}}\sigma^{\frac{1}{4}} K^{-\frac{11}{4}}, \tag{2.22}$$

where σ is the surface tension coefficient. Note that the magnitude of the exponent in (2.20)–(2.22) is smaller than 4 (corresponding to the Phillips' equilibrium spectrum).

It is important to note that the weak turbulence theory in its simplest form cannot explain the narrow angular distribution of wave energy for stationary ocean surface conditions. In this connection it should be noted that the interaction of a wave field and a non-potential mean surface current has recently been investigated theoretically. It was discovered that induced 'scattering' of surface waves on the shear current gives a narrow angular spectrum of gravity waves. But the mechanisms of formation of the angular wave spectra require additional investigations.

In summary, we can list the following important aspects of modern wave hydro-dynamic theory that present great interest for future ocean microwave remote sensing applications:

- Non-linear interactions of multi-scale surface waves.
- Generation of 2D and 3D surface wave structures.
- Modulation of short surface waves by long surface waves.
- Generation and evolution of the surface waves induced by non-uniform current field.
- Damping of surface waves due to turbulence.
- Development of surface wave instabilities and effects of wavenumber spectrum excitation.
- Non-linear dynamics and spatially temporal reorganization of subsurface hydrophysical fields including shear flows.
- Development of thermohaline convective processes in subsurface ocean layer.

2.4 WAVE-BREAKING AND FOAM/WHITECAPS STATISTICS

Wave-breaking is the main cause of two-phase non-uniformity occurring at the ocean–atmosphere interface. The nature of wave breaking is conditioned by a disruption of the equilibrium between the redistribution of energy into the wave spectrum on the one hand, and by atmospheric excitation (pumping) of wind waves in the range of the spectral maximum on the other hand. As a result of this redistribution, which occurs very slowly, wind-generated waves become unstable and break.

Several mechanisms of wave breaking are known, one of which is described as follows. The wave-breaking process starts when a surface wave has progressed to the stage where its amplitude elevates to some maximum value. This value is determined by equality between the shock-stalling speed and phase-speed of a surface wave. The sharpest crest wave angle of the limiting wave should be equal to 120°, and the maximum possible steepness of the breaking wave should be equal to $Ka = 2\pi a/\Lambda = 0.448$ [19]. Another mechanism affecting the process involves the effects of fluctuations of air flow over the ocean surface. The intensity of wave breaking depends on the presence of the surface wind drift and swell [20]. In this model the limiting wave height is less, and is estimated to be approximately 1/3 from the Stokes' wave-limiting configuration. However, the classical point of view on the wave-breaking mechanism associated with the appearance of Stokes' limiting wave height is not an adequate criterion. A set of laboratory and field observations and measurements shows that a surface wave breaks when its steepness exceeds approximately $Ka = 0.2–0.3$ and may sometimes be less. Breakers associated with small steepness are caused by long–short wave–wave interaction. Breakers associated with large steepness are caused by 2D or/and 3D wave instability [21, 22].

In the open ocean, the wave-breaking process begins earlier than is suggested by theory. In nature, the influence of surface current and wind speed fluctuations is significant. As a result, the conditions of large- and small-scale surface wave–wave interactions are changed, and the limiting-wave configurations are determined by dynamical parameters of the ocean boundary layer [23].

Geometry and structure of the wave-breaking zone can be illustrated by Figure 2.3. There are at least three types of ocean wave breaking: (1) plunging, (2) streaming, and (3) swelling. In the zone of wave breaking, the stationary turbulent flow of the two-phase mixture of water and air is formed analogously to the flow on a downhill surface.

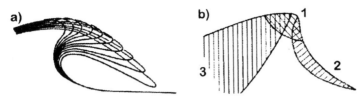

Figure 2.3. Temporary evolution of wave profile (a) and dynamical zones of wave breaking (b). 1—speed of water particles is more than phase-speed of a surface wave; 2—acceleration of water particles is more than the acceleration of gravity g; 3—acceleration of the water particles is less than $g/3$. (From Bunner & Peregrine [27].)

Hydrodynamic theory of two-phase flows in the ocean has not been fully developed, but some theoretical estimates give good agreement with laboratory measurements at early stages of the wave-breaking process [24–26].

Ocean wave-breaking processes are usually accompanied by the formation of two-phase dispersed structures at the air–water interface. They are an important electromagnetic factor that have a great effect on the microwave propagation characteristics of the ocean boundary layer. In particular, the physical properties and dispersed structure of foam/whitecaps were investigated in the laboratory and in nature [28–33].

The classification and some geometrical parameters of the ocean-dispersed structures are shown in Table 2.2. The data were collected from the literature and the author's own observations.

At very high wind speed conditions ($V > 15$ m/s), modulation instabilities and bifurcations of large gravity wind wave cause breaking and intensive foam/whitecaps processes. Under these conditions, both a transformation of low-frequency wavenumber spectrum and a change of area fraction, geometry and statistics of the wave-breaking field and foam/whitecaps coverage have occurred. In this view, the term 'wave-breaking field' means full hierarchy of oceanic two-phase dispersed structures that are formed at the air–water interface. As shown in Table 2.2, these structures essentially vary in composition and particle size.

Special investigations of wave spectrum dynamics and wave-breaking field characteristics (area statistics and size-distribution of foam/whitecaps) were made at the U.S.S.R. Space Research Institute using different types of nadir-viewing optical cameras [34–37]. A large number of aircraft experiments was conducted in the northwestern Pacific near the Kamchatka Peninsula coast (1981–1991). In particular, the high-resolution airspace MKF-6 six-band photo camera was used. The altitude of flight was changed from 300 m to 5000 m, so that high-resolution optical images of the ocean surface were obtained. Some digital algorithms for optical data processing were applied. They included standard two-dimensional Fast Fourier Transforms (FFTs) for the investigation of the spatial structure of large-scale wind waves, and special metrical procedures for investigation of foam/whitecaps geometry and statistics.

For the analysis of foam/whitecaps statistics, we used metrics of the following parameters: area A, perimeter P, and maximum and minimum linear size L_{max} and L_{min} for each object. To analyze the object's shape and topological shape attributes, simple

Table 2.2. Classification and parameters of the two-phase oceanic media

Main properties	Whitecaps	Foam streaks	Spray	Aerosol	Subsurface bubbles
Spatial size, m	0.5–10	3–30	10–20 (local clouds)	>100	>100
Thickness of layer, m	0.01–0.5 (mixture)	0.01–0.05 (monolayer)	0.2–1.5 and more	0.5–10	0.01–0.05
Volume water concentration, %	20–50	<5–10	0.01–0.1	<0.01	0.5–1.0
Size of particles, cm	0.5–1.0	0.01–0.5	0.01–0.1	<0.01	<0.01
Lifetime (stability)	Seconds (unstable)	Minutes, hours (stable)	Seconds (unstable)	Minutes (stable)	Hours (stable)

non-dimensional metrics such as the relations P^2/A and $A/(L_{max}L_{min})$ were calculated. Also, the fractal properties of optical binary images, i.e. images of 'background-objects' type, were investigated quantitatively. As a result, the foam/whitecaps statistical distributions were designed for fixed wind-wave-generated ocean conditions that change from still to storm due to strong wind fetch.

A strong transformation of foam/whitecaps statistics was observed due to wind fetch. Size-distribution, area-distribution, and distribution of geometrical parameters of the foam/whitecaps objects changed significantly. The dependence of the foam + whitecaps area fraction W on wind speed is shown using the well-known power-type law: $W(\%) = aV^b$, where a and b are empirical constants. The optimal values of the constants are: $a = 4.5 \cdot 10^{-4}$ and $b = 3.5$ in the range of wind speed $5 < V < 25$ m/s [38, 39]. But in fully developed storm conditions, a 'saturation' effect can take place. In this case, the value W does not depend on the ocean surface state, and the behavior of the function $W(V)$ is defined as the energetic balance in the wind-wave system.

New statistical characteristics of ocean dynamics are the fractal (multi-fractal) dimensions of the wave-breaking field and foam/whitecaps. Using airspace optical data and digital methods of two-dimensional image processing, it was found that the fractal dimension is a sensitive parameter of the ocean surface state. In remote sensing applications, the fractal dimension of the wave-breaking field can be considered as an alternative to the Beaufort wind force scale. Optical airborne measurements show that for foam streaks and whitecaps the value of the fractal dimension is different.

The fractal structure of the wave-breaking field and foam coverage was investigated in detail by using the method of air photography [40, 41]. The term 'fractal or multi-fractal dimension' has wide use in non-linear physics and dynamic system theory. This term was introduced recently as a description of self-similar sets of mathematical and natural objects [42] also in hydrodynamics and hydromechanics [43]. At present, the concept of fractal dimension is applied to the interpretation of geophysical remote sensing data. In particular, problems of scaling and multi-fractal analysis of the ocean surface are discussed with the use of microwave and optical data [44–46].

Let us consider some optical results. Optical images of an ocean–storm surface reflect non-linear dynamics of the wave-breaking process by visible radiance. Therefore, wave-breaking fields can be considered as fractal (multi-fractal) sets. Using this idea, two procedures were suggested for calculating fractal dimensions of a wave-breaking field. The first procedure, based on the 'box counting' method, was used for the analysis of small-scale optical images (the MKF-6 images, scale of imagery being 1:40 000). The fractal dimension or Hausdorff dimension D_H of a set is given by:

$$D_H = \lim_{r \to 0} \frac{\log N(r)}{\log \dfrac{1}{r}}, \tag{2.23}$$

where $N(r)$ is the smallest number of squares of side r required to completely cover the set (binary image, for example).

The second procedure was based on a simple relationship between area A and perimeter P of a single fractal object. The method was applied in the analysis of large-scale optical images (standard aerial pictures, scale of imagery being 1:3000). In the simplest case, the fractal dimension D_s of each visible object (foam streaks or whitecap) was estimated from the well-known area–perimeter relationship:

$$P \sim \left(\sqrt{A}\right)^{D_s}, \tag{2.24}$$

where A is the area and P the perimeter of a single object. The averaging of fractal dimension D_s at the object's ensemble on the image yielded the mean value of the fractal dimensions $\overline{D_s}$.

A set of optical images was processed using both digital algorithms. The results of fractal analysis are shown in Figure 2.4. For a moderately stormy ocean, it was found (Figure 2.4(a)) that the Hausdorff fractal dimension changes to within $D_H = 1.1$–1.3. At the same time the regression coefficients of the area–perimeter relationships give the value of the fractal dimension $\overline{D_s} = 1.39$ and 1.23 for the whitecaps and foam streaks respectively (Figure 2.4(b)).

Although there are many problems associated with the physical understanding of the wave-breaking phenomenon in the open ocean, manifestation of the scaling variability of the wave-breaking field permits the promising use of fractal dimensions as a top-level criterion of the ocean surface state.

An important aspect of the problem is the dynamics of wavenumber spectrum and wave breaking in the internal wave field. In general the influence of internal waves causes the wave-breaking wave statistics to change. Both the frequency of wave breaking and the total area fraction of foam coverage increase. The foam/whitecaps geometry will

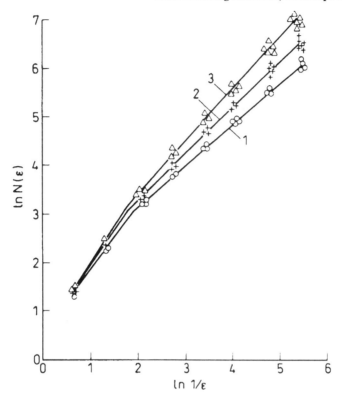

Figure 2.4(a). ln(N) vs. ln($1/\varepsilon$) for total foam coverage. Fractal dimension for three gradations of the Beaufort wind force: 1: 3–4 ($D_H = 1.05$); 2: 4–5 ($D_H = 1.15$); 3: 5–6 ($D_H = 1.25$).

also depend on the intensity and structure of the internal wave field. The field experiment showed that the fractal dimension of the wave-breaking field measured from the optical data varies in the presence or absence of an internal wave source. This effect was manifested during weak storm conditions. It is not only explained by the variance of wave-breaking frequency, but also by variance of the foam/whitecaps geometry.

Owing to the action of surface current non-uniformity, generation of hydrodynamic structures in the field of wave breaking is possible. The current gradients induce the speed of energy dissipation, and therefore change the regimes of wave breaking. Theoretically, different two-dimensional hydrodynamic concentric structures may appear in the field of a single internal wave or soliton. For example, the intensity of wave breaking increases in the zone of the current's convergence, and decreases in the zone of the current's divergence. Internal waves cause spatial modulation of limiting wind waves and accelerate the wave-breaking process. However, this question is not clearly understood.

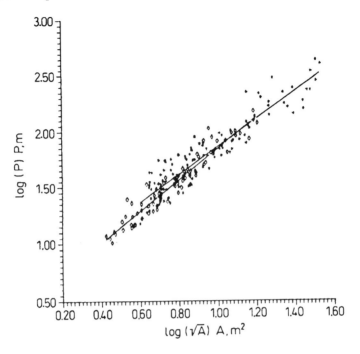

Figure 2.4(b). Log–log plots of perimeter (*P*) as a function of square root from area ($\sqrt{A}$) for foam streaks (*) and whitecaps ($\diamond$). Solid lines are linear least-square fits in different ranges of *A*. Beaufort wind force is 4.

REFERENCES

[1] Zakharov, V.E. (1968) 'Stability of periodic waves of finite amplitude on the surface of a deep water', *J. Appl. Mech. Tech. Phys.*, **2**, 190–194 (translated from Russian).

[2] Yuen, H.C. and Lake, B.M. (1982) 'Nonlinear dynamics of deep-water gravity waves', *Advances in Applied Mechanics*, Academic Press, Vol. 22, pp. 67–229.

[3] Craik, A.D.D. (1985) *Wave Interactions and Fluid Flows*. Cambridge: Cambridge University Press, 322 pp.

[4] Moiseev, S.S. and Sagdeev, R.Z. (1986) 'Problems of secondary instabilities in hydrodynamics and in plasma', *Radiophysics and Quantum Electronics*, **29**, No. 9, 808–812 (translated from Russian).

[5] Phillips, O.M. (1980) *The Dynamics of The Upper Ocean*. Cambridge: Cambridge University Press.

[6] Zakharov, V.E. and Zaslavskii, M.M. (1982) 'The kinetic equation and Kolmogorov spectra in the weak turbulence theory of wind waves', *Izvestiya, Atmospheric and Oceanic Physics*, **18**, No. 9, 747–753 (translated from Russian).

[7] Hasselman, K. (1962) 'On the nonlinear energy transfer in a gravity wave spectrum', *J. Fluid Mech.*, **12**, Part 3, 481–500.

[8] Mitsuyasu, H. and Honda, T. (1974) 'The high frequency spectrum of wind gener-
     ated waves', *J. Phys. Ocean.*, **30**, 185–195.

[9] Mitsuyasu, H. and Honda, T. (1982) 'Wind-induced growth of water waves', *J. Fluid Mech.*, **123**, 425–442.

[10] Keller, W.C., Plant, W.J., and Weissman, D.E. (1985) 'The dependence of X-band
     microwave sea return on atmospheric stability and sea state', *J. Geophys. Res.*, **90**,
     No. C1, 1019–1029.

[11] Donelan, M.A. and Pierson, W.J., Jr. (1987) 'Radar scattering and equilibrium
     range in wind-generated waves', *J. Geophys. Res.*, **92**, May, 4971–5029.

[12] Merzi, N. and Graft, W.H. (1985) 'Evaluation of the drag coefficient considering
     the effects of mobility of the roughness elements', *Annales Geophysicae*, **3**, No. 4,
     473–478.

[13] Apel, J.R. (1994) 'An improved model of the ocean surface wave vector spectrum
     and its effects on radar backscatter', *J. Geophys. Res.*, **99**, No. C8, 16269–16291.

[14] Monin, A.S. and Yaglom, A.M. (1971) *Statistical Fluid Mechanics; Mechanics in
     Turbulence.* The MIT Press, Cambridge, MA.

[15] Zaslavskiy, M.M. (1996) 'On the role of four-wave interactions in formation of
     space–time spectrum of surface waves', *Izvestiya, Atmospheric and Oceanic
     Physics*, **31**, No. 4, 522–528 (translated from Russian).

[16] Beal, R.C., Tilley, D.C., and Monaldo, F.M. (1983) 'Large- and small-scale evolu-
     tion of digitally processed ocean wave spectra from SEASAT synthetic aperture
     radar', *J. Geophys. Res.*, **88**, No. C3, 1761–1778.

[17] Volyak, K.I., Lyakhov, G.A., and Shugan, I.V. (1987) 'Surface wave interaction.
     Theory and capability of oceanic remote sensing'. *Oceanic Remote Sensing.* Nova
     Science Publishers, pp. 107–145 (translated from Russian).

[18] Gruishin, V.A., Raizer, V.Y., Smirnov, A.V., and Etkin, V.S. (1986) 'Observation
     of nonlinear interaction of gravitational waves by optical and radar techniques',
     Doklady of Russian Academy of Sciences, **290**, No. 2, 458–462 (in Russian).

[19] Lamb, H. (1932) *Hydrodynamics.* London: Cambridge University Press. (New
     York: Reprinted by Dover Publications, Inc., 1945.) 738 pp.

[20] Phillips, O.M. and Banner, M.L. (1974) 'Wave breaking in the presence of wind
     drift and swell', *J. Fluid Mech.*, **66**, No. 4, 625–640.

[21] Su, M.-Y. and Green, A.A. (1984) 'Coupled two- and three-dimensional instabili-
     ties of surface gravity waves', *Phys. Fluids*, **27**, No. 1, 2595–2597.

[22] Su, M.-Y. (1987) 'Deep-water wave breaking: experiments and field measure-
     ments', *Nonlinear Wave Interactions in Fluids. The winter annual meeting of the
     American Society of Mechanical Engineers.* Boston, MA, December 13–18. The
     American Society of Mechanical Engineers, N.Y., pp. 23–36.

[23] Kitaigorodskii, S.A. (1984) 'On the fluid dynamical theory of turbulent gas transfer
     across an air–sea interface in the presence of breaking wind-waves', *J. Phys.
     Oceanogr.*, **14**, 960–972.

[24] Longuet-Higgins, M.S. and Turner, J.S. (1974) 'An "entraining plume" model of a
     spilling breaker', *J. Fluid Mech.*, **63**, Part 1, 1–20.

[25] Hasselman, K. (1974) 'On spectral dissipation of ocean waves due to
     whitecapping', *Boundary-Layer Meteorology*, **6**, 107–127.

[26] *The Ocean Surface. Wave Breaking, Turbulent Mixing and Radio Probing* (1985). Y. Toba and H. Mitsuyasu (eds.). Dordrecht: D. Reidel.

[27] Bunner, M.L. and Peregrine, D.H. (1993) 'Wave breaking in deep water', *Annu. Rev. Fluid Mech.*, **25**, 373–397.

[28] Miyake, Y. and Abe, T. (1948) 'A study of the foaming of sea water. Part 1', *J. Marine Res.*, **7**, No. 2, 67–73.

[29] Abe, T. (1963) 'In situ formation of stable foam in sea water to cause salty wind damage', *Paper Meteorol. Geophs.*, **14**, No. 2, 93–108.

[30] Monahan, E.C. and Zietlow, C.R. (1969) 'Laboratory comparison of fresh-water and salt-water whitecaps', *J. Geophys. Res.*, **74**, No. 28, 6961–6966.

[31] Monahan, E.C. and MacNiocaill, G. (1986) *Oceanic Whitecaps*. Dordrecht, Holland: D. Reidel.

[32] Peltzer, R.D. and Griffin, O.M. (1988) 'Stability of a three-dimensional foam layer in sea water', *J. Geophys. Res.*, **93**, No. C9, 10804–10812.

[33] Raizer, V.Yu. and Sharkov, E.A. (1980) 'On the dispersed structure of sea foam', *Izvestiya, Atmospheric and Oceanic Physics*, **16**, No. 7, 548–550 (translated from Russian).

[34] Bondur, V.G. and Sharkov, E.A. (1982) 'Statistical properties of whitecaps on a rough sea', *Oceanology*, **22**, No. 3, 274–279 (translated from Russian).

[35] Raizer, V.Yu., Smirnov, A.V., and Etkin, V.S. (1990) 'Dynamics of the large-scale structure of the disturbed surface of the ocean from analysis of optical images', *Izvestiya, Atmospheric and Oceanic Physics*, **26**, No. 3, 199–205 (translated from Russian).

[36] Mityagina, M.I., Pungin, V.G., Smirnov, A.V., and Etkin, V.S. (1991) 'Changes of the energy-bearing region of the sea surface wave spectrum in an internal wave field based on remote observation data', *Izvestiya, Atmospheric and Oceanic Physics*, **27**, No. 11, 925–929 (translated from Russian).

[37] Raizer, V.Y. (1994) 'Wave spectrum and foam dynamics via remote sensing'. I.S.F. Jones, Y. Sugimori, and R.W. Stewart (eds.) *Satellite Remote Sensing of the Ocean Environment*. Japan: Seibutsu Kenkyusha, pp. 301–304.

[38] Monahan, E.C. and O'Muircheartaigh, I.G. (1986) 'Whitecaps and passive remote sensing', *Int. J. Remote Sensing*, **7**, No. 5, 627–642.

[39] Monahan, E.C. and O'Muircheartaigh, I.G. (1980) 'Optimal power-law description of oceanic whitecap coverage dependence on wind speed', *J. Phys. Oceanogr.*, **10**, No. 2, 2094–2099.

[40] Raizer, V.Yu. and Novikov, V.M. (1990) 'Fractal dimension of ocean-breaking waves from optical data', *Izvestiya, Atmospheric and Oceanic Physics*, **26**, No. 6, 491–494 (translated from Russian).

[41] Raizer, V.Y., Novikov, V.M., and Bocharova, T.Y. (1994) 'The geometrical and fractal properties of visible radiances associated with breaking waves in the ocean', *Annales Geophysicae*, No. 12, 1229–1233.

[42] Mandelbrot, B.B. (1983) *The Fractal Geometry of Nature*. New York: W.H. Freeman.

[43] Barenblatt, G.I. (1978) *Similarity, Self-similarity, Intermediate Asymptotics*. Leningrad: Gidrometeoizdat.

[44] Glasman, R.E. (1991) 'Statistical problems of wind-generated gravity waves arising in microwave remote sensing of surface winds', *IEEE Trans. Geosci. Remote Sensing*, **29**, No. 1, 135–142.

[45] Glasman, R.E. (1991) 'Fractal nature of surface geometry in a developed sea', *Nonlinear Variability in Geophysics*. D. Shertzer and S. Lovejoy (eds.), Dordrecht: Kluwer, pp. 217–226.

[46] Tessier, Y., Lovejoy, S., Schertzer, D., Lavallee, D., and Kerman, B. (1993) 'Universal multifractals indices for ocean surface at far-red wavelengths', *Geophys. Res. Lett.*, **20**, 1167–1170.

3

Microwave emission of the ocean

3.1 THE MAIN MECHANISMS

Microwave diagnostic techniques are based on analyses of the thermal emission of natural medium or/and objects in the range of $\lambda = 0.1–100$ cm radiowaves. The microwave frequencies are usually specified as the following standard ranges:

P-band	0.230–1.000 GHz
UHF-band	430–1300 MHz
L-band	1.530–2.700 GHz
S-band	2.700–3.500 GHz
C-band	3.700–4.200 GHz (downlink)
	5.925–6.425 GHz (uplink)
X-band	7.250–7.745 GHz (downlink)
	7.900–8.395 GHz (uplink)
Ku-band	10.700–18.000 GHz (has multiple acceptances)
Ka-band	18.000–37.000 GHz (has multiple acceptances)

Radiowaves are highly sensitive to the variations of the parameters and structure of environmental media which are connected with a larger depth of penetration in comparison to infrared or optical (visual) electromagnetic waves. Also, radiowaves are not easily absorbed in the atmosphere and clouds, and therefore permit more effective observations and monitoring of the ocean.

The effectiveness and reliability of microwave diagnostics depend upon the level of knowledge of the physical mechanisms of thermal emission from the medium being investigated. The scope of tasks included in studies of the ocean and atmosphere turn out to be sufficiently wide to encompass practically all basic issues of the theory of electromagnetic wave propagation. In general the theory is based on the solution of the Maxwell equations for lossy dielectric media including rough random surfaces. Many useful aspects of microwave remote sensing theory and practice are considered in the literature [1–4]; however, applications to ocean problems are very limited. Therefore, we develop a microwave concept that describes the oceans as random and non-homogeneous dynamical media with distributed multi-scale parameters. Anyway, we subdivide this into two main classes of oceanic non-uniformity: geometrical and volume.

The geometrical class relates to the ocean surface waves whose geometry and statistics are dependent on many hydrometeorological factors that, in term, determine the interaction between the atmosphere and ocean. In general, surface waves are presented as a non-stationary and non-uniform field of disturbance that has both deterministic and random components.

The volume class of non-uniformities is a two-phase (air–water) dispersed system, and is formed on the ocean (sub)surface as a result of the waves breaking or migration of gaseous bubbles from the deep water to the surface.

Geometrical non-uniformities are the integrated part of the ocean–atmosphere system. Dispersed structures of that system appear only under certain conditions. There are no universally accepted methods for the solution of electrodynamic problems for multi-scale and multi-component media as a non-uniform ocean boundary layer. Therefore, the design of a complete quantitative microwave model of the ocean–atmosphere system is extremely complex, because the total influence of all hydrophysical factors must be considered. However, it is possible to consider and investigate the microwave effects of each factor, taken separately, at least in the context of existing theoretical formulations and approaches.

The ocean microwave emission mechanisms due to the influence of geometrical factors include the following principal effects:

- Mirror reflection from a small-scale roughness surface.
- Diffused incoherent scattering on multi-scale surface irregularities.
- Coherent scattering from spatial–temporal correlating surface irregularities.
- Resonant scattering from surface irregularities with geometrical sizes which are comparable with the electromagnetic wavelength.
- Multiple scattering and shadows on large-scale irregularities.

The second class of ocean non-uniformities—volume (bubble–foam–spray)—has a completely different electrodynamic nature. The main mechanisms include: scattering, dissipation, and absorption of the microwave emission by separate particles, or by their aggregates like bubbles of foam and whitecaps, water drops, aerosols or gaseous bubble populations in the water.

Dielectric properties of salt water

The emissivity of the ocean depends on the dielectric properties of the water skin-layer. In the microwave range, the permittivity of ocean water is a complex function of the temperature and salinity, and also of the wavelength. Usually the Debye or more adequate Cole–Cole relaxation models [5] are used for microwave remote sensing applications. The latter gives a good approximation for the complex dielectric constant of salt ocean water:

$$\varepsilon_{\mathrm{w}} = \varepsilon_{\mathrm{w}}' - i\varepsilon_{\mathrm{w}}'' = \varepsilon_{\infty} + \frac{\varepsilon_{\mathrm{s}} - \varepsilon_{\infty}}{1 + \left(i\dfrac{\lambda_{\mathrm{s}}}{\lambda}\right)^{1-\alpha}} - i60\sigma\lambda, \tag{3.1}$$

where λ is the wavelength; σ is the conductivity of the water; ε_s is the low-frequency permittivity; ε_∞ is the high-frequency permittivity; λ_s is the critical wavelength which is related to the relaxation time τ by $\lambda_s = 2\pi c \tau$, and α is the Cole–Cole parameter which depends on the relaxation time distribution (usually $\alpha = 0.01$–0.30). At $\alpha = 0$, the Debye model is recovered from (3.1).

All parameters σ, ε_s, ε_∞, and λ_s are non-linear functions of the water temperature and salinity (or an aqueous NaCl solution). First calculations of the water permittivity [6, 7] showed strong dependencies on $\varepsilon_w(\lambda)$. More detailed estimations in a wide range of wavelengths $\lambda = 0.2$–200 cm [8–10] found strong dependencies of the permittivity on temperature and salinity. Effects of temperature and salinity on complex permittivity of the water are shown in Figures 3.1 and 3.2. These calculations were made using the Debye model. Other examples of the Cole–Cole diagrams and collected experimental data for salt water are presented in Figure 3.3. It is designed using the Cole–Cole model (3.1) when the wavelength λ is changed gradually.

The depth of the ocean skin-layer (depth of the electromagnetic wave's penetration) changes in relation to the water parameters. It can be calculated by the formula

$$l = (2\pi/\lambda)^{-1} \left[\frac{\varepsilon_w'}{2} \left(\sqrt{1 + \tan^2 \delta_w} - 1 \right) \right]^{-\frac{1}{2}} \tag{3.2}$$

where $\tan \delta_w = \varepsilon_w'' / \varepsilon_w'$ is the tangent of dielectric losses.

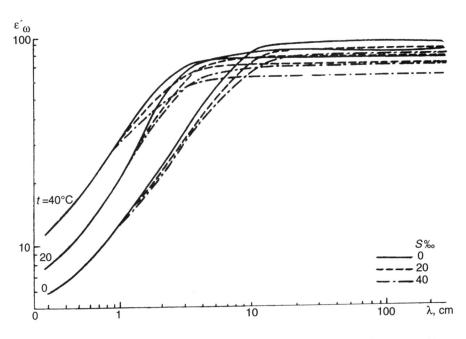

Figure   3.1. Real part of complex permittivity of sea water. Different values of temperature (t) and salinity (s) are denoted.

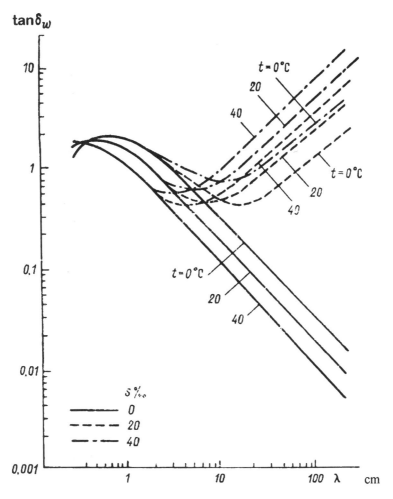

Figure 3.2. Tangent of dielectric losses of sea water. Different values of temperature (t) and salinity (s) are denoted.

The calculations show that the depth of penetration of millimeter and centimeter microwaves in the ocean water is equal to $l \approx (0.01–0.1)\lambda$, where λ is the wavelength in free space. In this range of wavelengths, value l weakly depends on the temperature and salinity of the water. But in the decimeter range, the depth of the skin-layer depends essentially on salinity and temperature, and can be equal to several centimeters (Figure 3.4). Thus, the optimal range of wavelengths for microwave remote measurements of sea surface temperature variations has to be $\lambda = 3–8$ cm, but for remote sensing of the ocean salinity field the range of wavelengths has to be $\lambda = 18–75$ cm.

As wavelength increases, the influence of the salinity on the brightness temperature of the ocean increases. At the same time, the accuracy of the airspace microwave

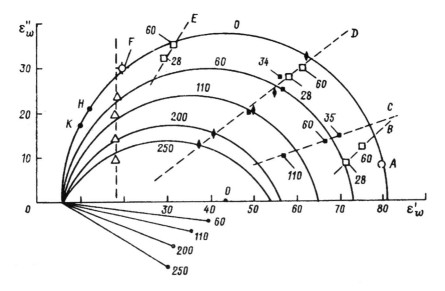

Figure 3.3. The Cole–Cole diagrams for NaCl water solutions. Temperature 20°C. Numbers are the values of salinity. Solid line—model calculation. Dashed line—measurements at the wavelengths: A—λ = 17.24 cm;   B—λ = 9.22 cm;   C—λ = 10 cm;   D—λ = 3.2 cm; E—λ = 1.26 cm; F—λ = 0.8 cm; H—λ = 0.5 cm; K—λ = 0.4 cm [10].

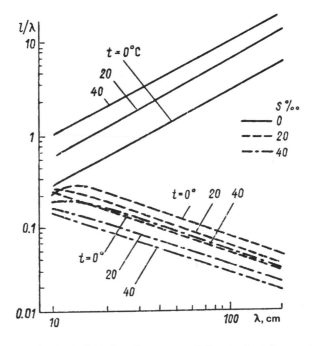

Figure 3.4. The relative depth of skin-layer for sea water. Different values of temperature (t) and salinity (s) are denoted.

measurements of salinity is not high, because the methods and algorithms used do not consider the effects of surface roughness and wave breaking [11, 12].

3.2 INFLUENCE OF SURFACE WAVES

Sea surface waves have been studied by research radiophysicists for more than 25 years. The first experiments and numerical estimates were made in the early seventies [13–15]. At first, radiophysical effects from the surface wave were evaluated using a one-scale and then a two-scale model. For example, a well-known two-scale model that used some combinations of the methods of geometrical optics and the theory of small perturbations was used to estimate the influence of large-scale and small-scale irregularities (roughness) on the scattering and emissivity of the sea surface [15]. Some results of these investigations are shown in Figure 3.5.

All previous calculations of the microwave emission of the sea surface were based on normal distributions of the height and slope of large sea waves. In general, the Gaussian function of the slope distribution or the Gram–Charlier series with standard deviations which linearly depend on the wind speed [16] was used. Calculations of small-scale roughness effects were conducted in the zero order of the method of small perturbations, with the application of modified reflection coefficients.

The application of the Gaussian law of wave slope distribution supposes that random surface irregularities can be represented by the statistical ensemble of linear flat waves. This is true if the long surface gravity waves on a deep water are considered, which

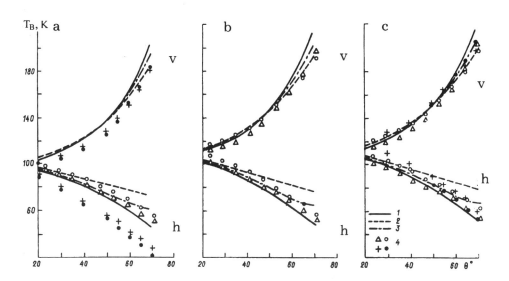

Figure 3.5. Dependencies of the sea surface brightness temperature on view angle. v—vertical polarization; h—horizontal polarization. The wavelength (frequency) of emission: (a) $\lambda = 21.4$ cm (1.4 GHz); (b) $\lambda = 7.5$ cm (4.0 GHz); (c) $\lambda = 2$ cm (7.5 GHz). Calculation: 1—flat water surface; the Kirchhoff model: 2—rms surface wave's slope 10°; 3—rms wave's slope 15°; 4—experimental data [15].

respond to the field of the spectral maximum of gravity wind waves. Actually, waves of different configurations and steepness are generated naturally and simultaneously in the ocean. The most common type of ocean surface waves in deep water is weakly non-linear gravity waves, and significant non-linear short gravity waves with finite amplitudes. The ensemble of weakly and strong non-linear waves cannot be described only by the Gaussian law of distributions.

The solution of practical problems of electromagnetic wave propagation across a rough (randomly corrugated) surface is generally based on the well-developed asymptotic methods of diffraction theory and theory of electromagnetic wave propagation [17–19]. For example, the methods of geometrical and physical optics, perturbation methods, and various numerical schemes are generally known. However, major difficulties are encountered in the solution of diffraction problems for close-packed irregularities of large steepness. The impedance approximations are not always satisfied under these conditions, but the more universal quasi-static (or macroscopic) approach can be used.

Resonant theory of the microwave emission of the small-scale rough water surface
This theory was developed to model microwave emission effects from gravity–capillary waves and capillary ripples on the ocean surface [20]. The method of small perturbations for the calculation of the intensities of two diffracted maximums and the mirror reflective component was applied. Analysis of the one-dimensional and two-dimensional (cylindrical) sinusoidal rough surface results in simple analytical relationships. In the last case the contrast of the brightness temperature with respect to the smooth surface is equal to:

$$\Delta T_B \approx T_0 \cdot (k_0 a)^2 \, G\left(\frac{K}{k_0}, \varepsilon, \theta, \varphi\right), \tag{3.3}$$

where $k_0 = 2\pi/\lambda$, $K = 2\pi/\Lambda$; Λ and a are the wavelength and the amplitude of sinusoidal irregularities; $G(\ldots)$ is the complex resonant function dependent on the dielectric constant of the water $\varepsilon(\lambda)$, the angle of view from the nadir θ, the azimuth angle φ, and polarization; T_0 is the thermodynamic temperature of the water surface.

With the constraint

$$2n\frac{K}{k_0}\sin\theta \cdot \cos\varphi + \left(n\frac{K}{k_0}\right)^2 = \cos^2\theta, \quad n = \pm 1, \pm 2,\ldots \tag{3.4}$$

resonant effects in the microwave emission of the rough surface appear [21, 22]. The best conditions for observing resonant maximums are realized at the nadir angle of view. This model provides the possibility of separating the influence of different polarizations on the microwave emission at the nadir, and at the grazing angles of view. Figures 3.6, 3.7 and 3.8 show some examples of calculations of the resonant function $G(\ldots)$. Parameters of the model are changed.

Numerical estimations show that the value of the brightness temperature contrast at the expense of the small surface sinusoidal irregularities influence can reach approximately $\Delta T_B \approx 30\,\text{K}$ at the resonant maximums. The effects were measured by

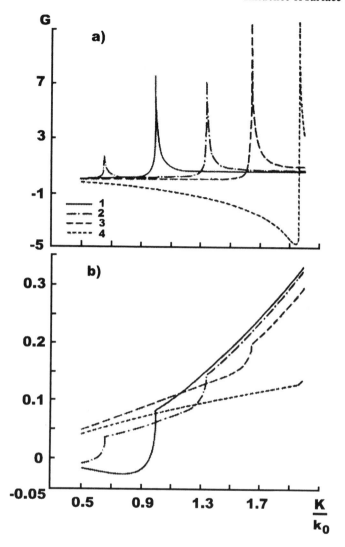

Figure 3.6. The resonant function $G(K/k_0)$ at wavelength $\lambda = 2$ cm for vertical (a) and horizontal (b) polarization. Azimuth angle $\varphi = 0°$. View angle is changed: $1—\theta = 0°$; $2—\theta = 20°$; $3—\theta = 40°$; $4—\theta = 70°$.

using different microwave radiometers in laboratory conditions, and good agreement between theory and experiment was shown. But in nature these effects were not recorded reliably.

Within the limits of the approach of small perturbations, extensions of the theory can be made in the case of the random roughness surface. For example, if the statistics of the random surface are determined by the two-dimensional wavenumber spectrum of the roughness, formula (3.3) takes the form:

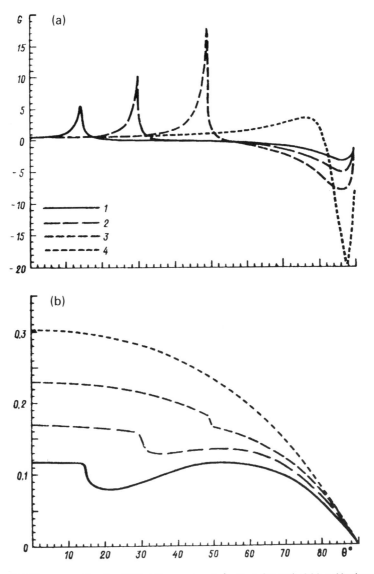

Figure 3.7. The resonant function $G(\theta)$ at the wavelength $\lambda = 2$ cm for vertical (a) and horizontal (b) polarization. Azimuth angle $\varphi = 0°$. 1—$K/k_0 = 1.25$; 2—$K/k_0 = 1.5$; 3—$K/k_0 = 1.75$; 4—$K/k_0 = 2.0$.

$$\Delta T_{\rm B} = 2T_0 k_0^2 \int_0^{2\pi} \int_{K_{\min}}^{\infty} G\left(\frac{K}{k_0}, \varepsilon, \theta, \varphi\right) \cdot F(K, \varphi) K \, {\rm d}K \, {\rm d}\varphi, \qquad (3.5)$$

where $F(K, \varphi)$ is the wavenumber spectrum of a rough surface. Usually the power wavenumber spectrum $F(K, \varphi) = AK^{-n} \cdot Q(\varphi)$ (where A and n are the parameters depending on the dynamical characteristics of the ocean boundary layer, and $Q(\varphi)$ is the

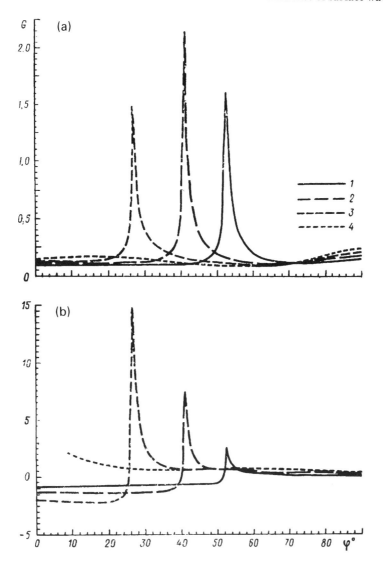

Figure 3.8. The resonant function $G(\varphi)$ at the wavelength $\lambda = 2$ cm for vertical (a) and horizontal (b) polarization. View angle $\theta = 70°$. $1—K/k_0 = 1.25$; $2—K/k_0 = 1.5$; $3—K/k_0 = 1.75$; $4—K/k_0 = 2.0$.

angular function) is used. Changes in the parameters A and n show an increase in some variations of the brightness temperature contrast. In practice, modeling was undertaken for $A = 0.01–0.0001$ and $n = 2–5$. The convolution (3.5) gives strong smoothing to the resonant maximums (Figure 3.9). The model was applied to the analysis of the ocean microwave data taken from the major part of field experiments and has shown good possibilities for radiospectroscopy of gravity–capillary surface waves in the ocean

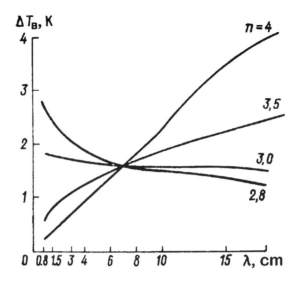

Figure 3.9. Brightness temperature contrast (at nadir view angle) of small-scale random water surface at different wavelengths of microwave emission. Model (3.5) with the spectrum: $F(K, \varphi) = AK^{-n} \cdot Q(\varphi)$; $A = 0.0002$; $Q(\varphi) = (1 + 0.5 \cos 2\varphi)/2\pi$. Value of the power exponent is changed: $n = 2.8$; 3.0; 3.5; 4 (denoted).

[21–23]. The model permits an explanation of the effects of anisotropy of ocean microwave emission at different polarizations, which sometimes was observed in nature using airborne radiometry [24].

Two-scale and three-scale modified models

Development of the microwave emission theory is connected with the modification of a two-scale model. The modification accounts for the influence of statistics of large-scale components of wind waves, and simultaneously the resonant features of the small-scale components. Also the model describes the effects of the polarization of microwave emission. Mathematical relations between the brightness temperature contrast using geometrical optics' approaches for the large-scale components, and the model (3.5) for small-scale components are:

$$\Delta \overline{T}_B = \int_{-\infty}^{+\infty} \int_{-\infty}^{+\infty} \Delta T_B(z_x, z_y) P(z_x, z_y) \, \mathrm{d}z_x \, \mathrm{d}z_y, \tag{3.6}$$

where $P(z_x, z_y)$ is the probability distribution of wave slopes (in local coordinate system).

However, recently it was shown theoretically [25, 26] that diffraction of the incident electromagnetic field on large-scale components of surface irregularities cannot be considered by using Kirchhoff approximation at low grazing ($\theta > 70°$ from nadir) view angles. In this case the curvature of flat facets becomes an important parameter in a two-scale surface model.

Comparison of resonant theory and Kirchhoff approximation for thermal microwave emission from rough sea surface was made using small-scale as well as small-slope expansion techniques [27]. It was found in particular that the methods of calculation are identical. So, well-known small-perturbation approximation can be used even for large-scale rough surface with small slopes, and it is possible to correctly calculate the emissivity on the basis of relationship (3.5). In this case the value of the cutoff K_{min}, in principle, may be shifted to a more low-frequency range, which corresponds to an ensemble of large-scale surface waves with small slopes. However, at the grazing view angles one needs to take high-order terms in the small-slope approximation into account. These describe the effects of multiple scattering and shadows.

The next step in applying the ocean microwave emission theory includes the development of a three-scale model that considers: the statistical ensemble of gravity waves + resonant gravity–capillary waves + small-scale turbulent roughness or short waves with large steepness. This model is useful for microwave diagnostics of the ocean surface phenomena at long wavelengths ($\lambda = 8$–30 cm) at large antenna footprint averaging. It is essential that the dependence of the effective permittivity from geometry and statistics of small-scale random roughness be considered in developing the model. For example, using the quasi-static approach, the function $\varepsilon\{F(K)\}$, where ε is the effective permittivity of the statistically rough air–water interface, can be introduced into the two-scale model (3.5), (3.6). After that a variant of the three-scale model for the rough ocean surface will be obtained. But the numerical analysis of the three-scale model and its application to ocean remote sensing have not been carried out yet.

Contribution of short gravity waves

This model was developed to describe the microwave effects from a statistical ensemble of non-linear short gravity waves on a deep water or from the surface waves of the finite amplitude. This is a one-scale model which is based on geometric optics approximations. The special function of the slope distributions of non-linear surface waves ensemble is applied. Let us consider a multi-mode model of the rough surface in the general form

$$\xi(x) = \sum_{n=1}^{N} a_n \cos(K_n x + \psi_n) \tag{3.7}$$

where a_n and K_n are the amplitude constants and spatial frequencies of the harmonics or modes, ψ_n are the random phases, which are uniformly distributed over the interval $[0, 2\pi]$, and n is the number of surface modes. To determine the emissivity of the multi-mode random surface we must find the probability density of the distribution of the derivatives (surface slopes) of this process. In the case of non-synchronized phase the desired density is

$$P(z) = \frac{1}{2\pi} \int_{-\infty}^{+\infty} \prod_{n=1}^{N} J_0(Uz_n) \, e^{-iUz} \, dU, \tag{3.8}$$

where $J_0(x)$ is the zeroth-order Bessel function of the real argument and $z_n = K_n a_n = nC_n(Ka)^n$, where Ka is the initial wave steepness. The non-linearity is introduced via the coefficients a_n of the Stokes' expansion. For harmonics $N = 1, 2, 3, 4$ the distribution differs significantly from the Gaussian distribution, and begins to approach it only for $N \geqslant 5$ (Figure 3.10).

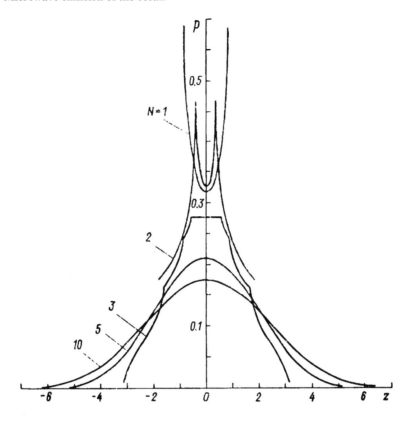

Figure 3.10. Transformation of slope distribution density for multi-mode random surface. Number of mode is changed: $N = 1; 2; 3; 5; 10$. Initial wave steepness $Ka = 0.75$.

The emissivity of the random multi-mode one-dimensional surface is now found by the averaging procedure:

$$\kappa(\theta) = 1 - \int_{-\infty}^{+\infty} |r(\cos \chi)|^2 P(\theta, z) \, dz, \tag{3.9}$$

where r is the Fresnel reflection coefficient for the vertical or horizontal polarization; χ is a local angle of incidence, which depends on the angle of view θ and the slope of the surface wave z. The brightness temperature contrast of the disturbed surface waves is equal to $\Delta T_B = (\kappa - \kappa_0)T_0$, where κ_0 is the emissivity of the smooth water surface and T_0 is the thermodynamic temperature of the water. Possible values of the brightness contrast for a random sinusoidal large-scale water surface are shown in Figure 3.11.

Microwave measurements of short non-linear gravity waves on the water surface were made in a tank using highly sensitive radiometers (on the superconducting Josephson detector) at the wavelengths $\lambda = 0.8$ and 1.5 cm [28–30]. Variations in the brightness temperature due to change in amplitude and wavelength of the surface waves were investigated. Examples of the modeling and collected experimental data are shown in Figures 3.12 and 3.13.

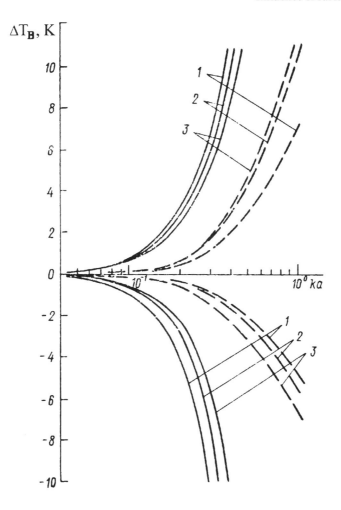

Figure 3.11. Dependencies of brightness temperature contrast on steepness of sinusoidal waves. Calculation by using the geometric optics approach. Wavelength of microwave emission is changed: $1—\lambda = 0.3$ cm; $2—\lambda = 0.8$ cm; $3—\lambda = 1.5$ cm. Different view angles: solid line—$\theta = 50°$; dashed line—$\theta = 30°$.

The strong dependence of the brightness temperature contrast on the steepness of the surface waves is manifest. Also the theoretical and the experimental data demonstrate a high sensitivity of microwave emission to the geometry and non-linearity of short gravity waves on the deep water. A comparison of theoretical and experimental data shows that the minimal value of the brightness contrast manifested due to the influence of weakly non-linear surface waves was $\Delta T_B = 0.2$ K. When the steepness of the surface waves increased, the brightness contrast reached a value of 8–10 K at both wavelengths of emission. Of special significance is the dependence of the ΔT_B on the parameter of non-linearity N (it is the number of harmonics which form the surface wave non-linear profile).

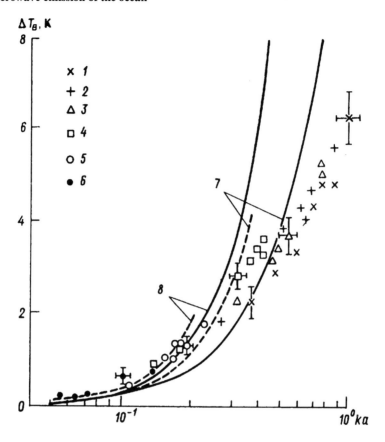

Figure 3.12. Dependencies of brightness temperature contrast on wave's steepness at the wavelength of emission $\lambda = 0.5$ cm; $\theta = 30°$, vertical polarization. Experimental data for the following wavelengths of the surface waves: 1—$\Lambda = 8$ cm; 2—$\Lambda = 10$ cm; 3—$\Lambda = 12$ cm; 4—$\Lambda = 18$ cm; 5—$\Lambda = 25$ cm; 6—$\Lambda = 40$ cm. Calculation: 7—one-mode approximation ($N = 1$); 8—double-mode approximation ($N = 2$); solid line—data without atmosphere contribution; dashed line—data with atmosphere contribution (standard model).

Similar calculations of the microwave emission can be made on the assumption of standard Gaussian wave slope statistics using a one-scale model of the sea surface. In this model the mean-square slopes of the surface waves depend linearly on the wind speed according to the well-known Cox & Munk approximation. The mean value of the gravity wave's slopes is of the order of 0.001–0.01. However, in the case of finite-amplitude surface waves this value may be 0.5–1, and the microwave effects due to the wave's non-linearity is more important than that due to Gaussian slope distribution. Experimental observations in an open water channel [31] have demonstrated the existence of steady three-dimensional symmetric water wave patterns. These surface waves were the result of bifurcation of uniform Stokes' waves with a steepness of 0.25 or more. Therefore, a more reliable interpretation of microwave data in the range of steepness 0.5–1.0 should be made with the use of a two- or three-dimensional multi-mode model, for example:

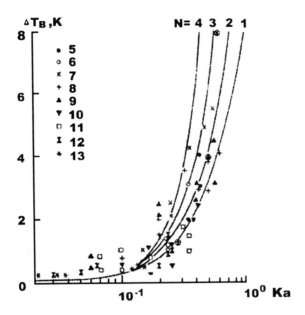

Figure 3.13. Dependencies of brightness temperature contrast on wave's steepness at the wavelength of emission λ =1.5 cm; θ =30°, vertical polarization. Solid line—calculation, figures near the curves are the values of N = 1, 2, 3, 4. Experimental data for the following wavelengths of the surface waves: 5—Λ = 1.3 cm; 6—Λ = 12.9 cm; 7—Λ = 15.4 cm; 8—Λ = 17.4 cm; 9—Λ = 20.7 cm; 10—Λ = 25 cm; 11—Λ = 32.2 cm; 12—Λ = 38.9 cm; 13—Λ =53.2 cm.

$$\xi(x,y) = \sum_{n}\sum_{m} A_{n,m}(K_x, K_y)\cos(nK_x x + \varphi_n)\cos(mK_y y), \tag{3.10}$$

where $A_{n,m}$ are the amplitudes of the modes: K_x and K_y are the wave numbers. In this case, the cross-polarized microwave effects due to non-linearity of large-scale surface waves can be observed.

A multi-mode ocean surface model which takes non-Gaussian statistics of wind wave slopes and the bifurcation's phenomena into account can be applied to the microwave diagnostics of short non-linear gravity waves and/or other surface disturbances with non-linear characteristics.

The quasi-static model of random rough surface
The idea is to model the surface roughness by a transition dielectric layer with effective parameters. The field of the air–water rough interface is described by a two-phase (air–water) plane-parallel layer, and the problem is reduced to the calculation of the spectral reflectivity and emissivity of the non-homogeneous macroscopic layer located on the water surface. First of all, simple cases of a regular rough surface with sinusoidal or rectangular profiles can be analyzed. The calculation technique reveals the sharp dependence of emissivity on the amplitude and geometry of the irregularities. It may be supposed that the emissivity increases significantly with the growth of the amplitude of

the irregularities owing to the influence of the transition layer. In addition the strong sensitivity of the emissivity to polarization of the electromagnetic field (or orientation of irregularities) can be manifested.

The common limits of the macroscopic approach are: $k_0 \ll K$, $k_0 a \ll 1$, where $K = 2\pi/\Lambda$, $k_0 = 2\pi/\lambda$, Λ and a are the horizontal and vertical dimensions of the irregularities, and λ is the radio wavelength. The application of the macroscopic theory for ocean remote sensing is justified by the inadequacy of the perturbation method, and impedance approaches for the analysis of steep and closely spaced irregularities. In the case when the skin-layer of the dielectric medium (or the depth of radio wave penetration into the medium) is comparable to the curvature radius of the rough surface, the application of these methods is not possible. The quasi-static model places no constraints on the parameter of steepness $Ka > 1$. This allows for a description of the microwave effect of very small-scale capillary waves having the form of cusped troughs (inverted wave) or turbulent surface deviations. The model is also evidently suitable for evaluating the microwave emission contribution of hydrophysical phenomena associated with macroscopic discontinuities of the ocean surface.

Let us suppose that microwave properties of the 'transition layer' depend on the volume concentration of the phase components of the mixture, geometry of the irregularities, relative orientation of the wavenumber vector $\mathbf{K}$ of the irregularities, and the electric vector $\mathbf{E}$ of the external electromagnetic field. The cases of $\mathbf{K} \| \mathbf{E}$ and $\mathbf{K} \perp \mathbf{E}$ correspond to vertical and horizontal polarization. The first calculations and laboratory radiometric measurements at a wavelength of $\lambda = 18$ cm showed a good sensitivity of the emissivity to amplitude variations, and configuration of the surface irregularities. This effect was tested in the laboratory with the use of a microwave radiometer at a wavelength of $\lambda = 18$ cm [32]. Surface irregularities on the water surface were created in the laboratory tank with the aid of a sheet of a radio translucent foam plastic with sinusoidal (or rectangular) profiles. The corrugated side of the sheet was pushed into the water with the smooth side turned to the antenna of the radiometer. In this manner the 'frozen' regular structure with the various and well-known parameters and geometry could be reproduced. Comparison of the tank test results showed good agreement between theoretical and experimental data. But application of the method of the interpretation of natural microwave data requires the following development of the quasi-static model and consideration in a more complex case of an ocean-like random surface. It was also found that the emissivity is higher for $\mathbf{K} \| \mathbf{E}$ than for $\mathbf{K} \perp \mathbf{E}$. The effect of strong 'polarization anisotropy' of the microwave emission was observed.

A comparison of the two approaches, 'resonance' and 'macroscopic', improves the limits of applicability for each microwave model by the parameter of wave steepness Ka. Microwave emission from the deterministic sinusoidal surface $z = a \cos Kx$ is considered as a test example.

Figure 3.14 shows the dependencies of the brightness temperature contrast on the steepness of the sinusoidal water surface $\Delta T_B(Ka)$ at the wavelength $\lambda = 18$ cm for fixed values $k_0 a = 0.05$, 0.10 and 0.15. Three different electrodynamic methods are used for the calculation: (1) the resonance model (theory of critical phenomena); (2) quasi-static (macroscopic) model; and (3) numerical solution of the diffraction problem based on the method of integral equations [33].

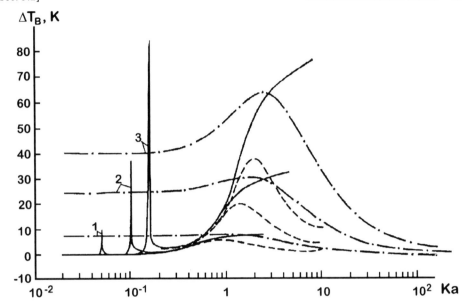

Figure 3.14. Theoretical graph of the brightness temperature contrast $\Delta T_B(Ka)$ for a sinusoidal water surface. $\lambda = 18$ cm; $\theta = 0$; $\varphi = 0$; surface temperature $T_0 = 293$ K. Complex dielectric constant of the water $\varepsilon_w(\lambda) = 71.1 + \iota 52.0$. Solid line—by resonance model; dashed line—by macroscopic model. Dashed line—by numerical method (calculations were made by V. Irisov). Parameter $k_0 a$: 1—0.05, 2—0.10, 3—0.15.

Thus, in the case of the resonance model, sharp maxima occur when the condition $K/k_0 = 1$ is satisfied (or for corresponding values of $Ka = k_0 a$). In the case of the macroscopic model, the graphs of $\Delta T_B(Ka)$ are more monotonic in the region $Ka < 0.5$. In the resonance model, the contrast ΔT_B is determined solely by the value of the parameter $k_0 a$, and is practically independent of the steepness Ka. In the region of steepness $Ka > 1$, the method of small perturbations is unsuitable and a numerical solution of the diffraction problem gives another maximum of the microwave emission. A similar path of the curves $\Delta T_B(Ka)$ is also observed in the case of the macroscopic model. Here, the brightness contrast tends asymptotically to zero ($\Delta T_B \to 0$) as $Ka \to \infty$ (actually for $Ka > 100$), which corresponds to the case of close-packed and steep irregularities. So, the microwave properties of such rough surfaces are adequate for the continuous dielectric medium.

However, in nature, the range of variation of the parameters Ka and $k_0 a$ may be quite broad, and may cross the limits of applicability of both models. Thus, it is advisable to apply both theoretical approaches to analyze ocean remote sensing data.

The quasi-static theory can be elaborated for a random rough surface with the continuous two-dimensional wavenumber spectra of small-scale irregularities [34]. A significant element of the model discussed below is the recourse to elements of the theory of random processes and fields [35]. Application of the theory permits circumvention of the difficulties encountered in the solution of the corresponding boundary-conditions electrodynamic problem for a random rough surface, and to apply the requisite statistical operations for ensemble of effective macroscopic parameters of the transition layer.

For microwave investigations, the numerical quasi-static model can be applied at wavelengths $\lambda = 3$–8 cm or longer. The following conditions of applicability of the quasi-static model can be formulated.

Let the random surface $z = \xi(x, y)$ be described by the spatial spectrum of roughness $S(K)$. The radius of curvature of the surface is estimated as $R \sim 1/(K^2 a)$; and the thickness of the skin-layer is estimated as $L \sim 1/(k_0 \sqrt{\varepsilon})$, where ε is the complex permittivity of the medium. Thus, the relation $R \sim L$, together with general conditions of macroscopic theory $k_0 a \ll 1$, $k_0 \ll K$, give the required limits:

$$a^2 \sim \overline{\xi^2} = \int_{K_{\min}}^{\infty} S(K) \, dK \ll k_0^2, \quad k_0 \ll K_{\min}, \tag{3.11}$$

$$(Ka)^2 \sim \overline{(\nabla \xi)^2} = \int_{K_{\min}}^{\infty} S(K) K^2 \, dK \ll \sqrt{|\varepsilon|}. \tag{3.12}$$

At wavelengths $\lambda > 5$–8 cm the permittivity of the salt water is about $|\varepsilon| \approx 30$–80, so that inequalities (3.11) and (3.12) are satisfied for any surface wave steepness $Ka \geqslant 1$. Therefore, the application of the quasi-static approach in this microwave range is well founded.

Let a stationary, spatially homogeneous perturbation field $z = \xi(x, y)$ be given with correlation function $B(r_2 - r_1) = B(\Delta x, \Delta y)$, or spectral density

$$S(K_x, K_y) = \int_{-\infty}^{+\infty} \int_{-\infty}^{+\infty} B(\Delta x, \Delta y) e^{-i(K_x \Delta x + K_y \Delta y)} \, d\Delta x \, d\Delta y. \tag{3.13}$$

The rough surface is interpreted within the framework of the probabilistic macroscopic approach as a realization of the random field $z = \xi(x, y)$, bounded by the domain $x, y \in [0, T]$. The interface between the two media is modeled by a continuous transition layer with an effective profile of the dielectric constant $\varepsilon_f(z)$ (Figure 3.15).

The depth profile $\varepsilon_f(z)$ must describe the geometrical and statistical singularities of the surface; its determination poses the basic problem of the macroscopic theory of rough

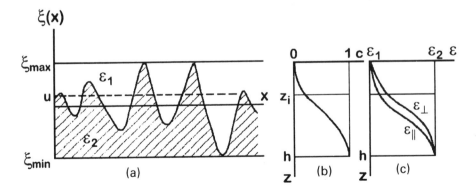

Figure 3.15. The diagram and parameters of the quasi-static model. 1—air; 2—water; (a) random rough surface and transition two-phase layer; (b) profile of the water concentration in the layer; (c) profile of the effective permittivity in the layer.

surfaces. The transition layer encompasses the interface and is a two-phase medium. We denote volume concentration of the underlying medium in the layer by c. This parameter, or more precisely its profile $c(z)$, specifically characterizes the distribution of the phases in the layer, which depends on the geometry of the surface. By knowing $c(z)$, we can find the effective profile $\varepsilon_f(z)$ from the equations of macroscopic electrodynamics, and then solve the problem of reflection and radiation from an inhomogeneous half-space. Since the field $z = \xi(x, y)$ is random, c and $\varepsilon_f(z)$ are random functions of the field.

Expressions for $\varepsilon_f(z)$ are derived on the basis of the solution of Maxwell's constitutive equations, one of which interrelates the electric displacement $\mathbf{D}$ and the electric field $\mathbf{E}$ in a linear dielectric medium: $\langle \mathbf{D} \rangle = \varepsilon_f \langle \mathbf{E} \rangle$. The solution of this equation is trivial in the quasi-static approximation

$$\varepsilon_f = \frac{(1-c)\varepsilon_1 E_1 + c\varepsilon_2 E_2}{(1-c)E_1 + cE_2} = \frac{(1-c)\varepsilon_1 + c\varepsilon_2 q}{1-c+cq}, \tag{3.14}$$

where $q = E_2/E_1$. The fields E_1 and E_2 are coupled through two-sided boundary conditions at the interface $z = \xi(x, y)$:

$$(\mathbf{N}, \varepsilon_1 \mathbf{E}_1) = (\mathbf{N}, \varepsilon_2 \mathbf{E}_2), \quad [\mathbf{N}, \mathbf{E}_1] = [\mathbf{N}, \mathbf{E}_2], \tag{3.15}$$

$$\mathbf{N} = \frac{1}{\gamma_0}\left\{ -\frac{\partial \xi}{\partial x}, \quad -\frac{\partial \xi}{\partial y}, \quad 1 \right\}, \quad \gamma_0 = \left(1 + \gamma_x^2 + \gamma_y^2\right)^{\frac{1}{2}},$$

where ε_1 and ε_2 are the dielectric constants of bounded media (air and water), and $\mathbf{N}$ is the unit normal to the rough surface.

One solution of the equations (3.14), and (3.15), assuming that $E_1 = E_0$ in the zeroth approximation (not taking into account the contribution of the depolarizing field), gives the following relationship for the effective permittivity of the transition layer:

$$\varepsilon_f = \varepsilon_1 + (\varepsilon_2 - \varepsilon_1) \sum_{n=1}^{\infty} q(1-q)^{n-1} c^n, \tag{3.16}$$

$$q = \left(1 - \alpha_0^2 \eta^2 \gamma_0^{-2}\right)^{\frac{1}{2}}, \quad \eta^2 = \frac{\varepsilon_2^2 - \varepsilon_1^2}{\varepsilon_2^2},$$

$$\alpha_0 = \begin{cases} \sin\theta + \gamma_x \cos\varphi \cos\theta + \gamma_y \sin\varphi \sin\theta, & \mathbf{K} \| \mathbf{E} \\ \gamma_y \cos\varphi - \gamma_x \sin\varphi, & \mathbf{K} \perp \mathbf{E} \end{cases}.$$

Since the random variable ε_f is the non-linear function of the random variable c, the series expansion of equation (3.16) in powers of c^n is useful in subsequent statistical operations.

To calculate the mean value and variance of the random variable ε_f, it is necessary to determine the corresponding moments of the random variable c. At this point we applied a new procedure that is based on principles of the statistical theory of excursions of random fields.

The inhomogeneous half-space with vertical distribution of the permittivity $\varepsilon_f(z)$ is represented by the multi-layer system [32]. Thus, the function $c(u)$ in an elementary layer of thickness Δu is the ratio of the total volume of all domains, bounded by levels u and $u + \Delta u$, to the total volume of the layer of the horizontal size T, i.e.

$$c(u) = \frac{V_T(u) - V_T(u + \Delta u)}{\Delta u T^2}. \tag{3.17}$$

The full volume of random field excursions above a fixed horizontal level is described by the Z_0-characteristic:

$$Z_0 = \frac{1}{T^2} \int_0^T \int_0^T g(x, y, u) \, dx \, dy, \quad g(x, y, u) = \begin{cases} 1, & \xi > u \\ 0, & \xi \leqslant u \end{cases}, \tag{3.18}$$

where $g(x, y, u)$ is the indicator function.
  In the limit $\Delta u \to 0$ we obtain

$$c(u) = \lim_{\Delta u \to 0} \frac{1}{\Delta u} \int_u^{u+\Delta u} Z_0(u_1) \, du_1 = Z_0(u). \tag{3.19}$$

The calculation of the first and second moments of the random variable c for the case of the uniform Gaussian field and large size T gives a general relation

$$M_c = \bar{c} = 1 - \Phi\left(\frac{u}{\sigma_\xi}\right), \tag{3.20}$$

$$D_c = \overline{c^2} - \bar{c}^2 = \frac{4}{T^2} \sum_{n=1}^{\infty} \frac{1}{n! \sigma_\xi^{2n}} \left[\Phi^{(n)}\left(\frac{u}{\sigma_\xi}\right)\right]^2 \int_0^\infty \int_0^\infty B^n(r_2 - r_1) \, dr_1 \, dr_2, \tag{3.21}$$

where $\Phi^{(n)}(x)$ is the derivative of the probability integral, and σ_ξ^2 is the full variance of the random field $z = \xi(x, y)$.
  The variance of the excursion volume is determined entirely by the form of the correlation function $B(r_2 - r_1)$ of the random field $z = \xi(x, y)$, and by the level u. The factorial in (3.21) makes the series converge rapidly.
  For simplicity, only the first and second terms of the series (3.16) can be used. Then the mean and variance of the random effective permittivity ε_f at the fixed level $z = u$ will equal

$$M_\varepsilon = \bar{\varepsilon}_f = \varepsilon_1 + (\varepsilon_2 - \varepsilon_1)\left[M_q M_c + M_q(1 - M_q)(M_q^2 + D_c)\right], \tag{3.22}$$

$$D_\varepsilon = \overline{\varepsilon_f^2} - \bar{\varepsilon}_f^2 = (\varepsilon_2 - \varepsilon_1)^2 \left(D_q D_c + M_c^2 D_q + M_q^2 D_c\right), \tag{3.23}$$

where M_q and D_q are the assembly average and variance of the random variable $q(\gamma_x, \gamma_y)$ (terms of q and c are assumed to be statistically independent in the averaging operation). Thus, the effective parameters of the transition layer, such as the profiles of $M_\varepsilon(z)$ and $D_\varepsilon(z)$, are well determined analytically.
  The quasi-static approximation can be used to assess the influence of small-scale random disturbances, such as short capillary waves, ripples and turbulent flutters,

associated with the interaction of air flows and rough surfaces. It is practically impossible to predict the geometry of disturbances in these situations. The contribution of the disturbances to the ocean microwave emission can be especially significant in a long-centimeter and decimeter wavelength range. In this region, the thickness of the skin-layer of salt water is greater than the characteristic scales of the surface disturbances, and the disturbances themselves are much smaller than the radiation wavelength. These conditions correspond to (3.11), and (3.12).

As an example, let us consider the microwave effects from the rough ocean surface with a power-type spectrum in a general form:

$$S(K_x, K_y) = \frac{4A\alpha\beta}{\left(\alpha^2 + K_x^2\right)^\nu \left(\beta^2 + K_y^2\right)^\mu}, \tag{3.24}$$

where A, α, β, ν, μ are constants. The spectrum (3.24) describes an anisotropic field of small-scale irregularities. For $\nu = \mu = 1$, this corresponds to an exponential correlation function:

$$B(\Delta x, \Delta y) = A \exp\left(-\alpha|\Delta x| - \beta|\Delta y|\right). \tag{3.25}$$

The full variance of a random field $z = \xi(x, y)$ is

$$\sigma_\xi^2 = \frac{1}{(2\pi)^2} \iint S(K_x, K_y)\, dK_x\, dK_y.$$

The moments of the roughness spectrum are

$$m_{ij} = \int_{\Delta K_x} \int_{\Delta K_y} S(K_x, K_y) K_x^i K_y^j\, dK_x\, dK_y. \tag{3.26}$$

The moments of the angular factor q in (3.16) are

$$\overline{q^i} = \iint q^i(\gamma_x, \gamma_y) p(\gamma_x, \gamma_y)\, d\gamma_x\, d\gamma_y, \quad i = 1, 2, \tag{3.27}$$

where $p(\gamma_x, \gamma_y)$ is the slope distribution of the random field $\xi(x, y)$. The distribution is assumed to be the Gaussian function:

$$p(\gamma_x, \gamma_y) = \frac{1}{2\pi\Delta^{1/2}} \exp\left[-\frac{m_{02}\gamma_x^2 - 2m_{11}\gamma_x\gamma_y + m_{20}\gamma_y^2}{2\Delta}\right], \tag{3.28}$$

$$\Delta = m_{20}m_{02} - m_{11}^2.$$

Consequently, the effective properties of the transition layer depend on the moments of the wavenumber spectrum m_{ij}, and the orientation angles θ and φ. Using the procedure for calculating the coefficients of reflection r_f and emission κ_f of a multi-layer structure with an arbitrary complex profile $\varepsilon_f(z)$, the brightness temperature of the rough surface T_B is determined:

$$r_f = F\{\varepsilon_f(z)\}; \quad \kappa_f = 1 - |r_f|^2; \quad T_B = \kappa_f T_0, \tag{3.29}$$

where F is the 'reflectivity operator' of the multi-layer system. It is important to note that the solution method used takes into account implicitly, the 'diffraction' properties of the transition layer over parameter h/λ, where h is its total thickness (Figure 3.15). The variance of the total thickness is $h \sim \sqrt{\sigma_\xi^2} = \sqrt{m_{00}}$.

The special numerical modeling was made using relations (3.22)–(3.29), with the spectrum of roughness (3.24) and correlation function (3.25). Parameters of the model varied over a wide range. Some results of the calculations are shown in Figures 3.16–3.20.

In the quasi-static model, the selection of the wavenumber spectrum's characteristics is very important. They are the spatial-frequency bands ΔK_x, ΔK_y, the value of the low-frequency cutoff K_{min}, and the exponents μ, v. These parameters cause the limits of applicability of the quasi-static model. Variations of the parameters v, μ and $\Delta K_{x,y}$ induce comparable variations in brightness temperature of the rough surface. Figures 3.18 and 3.19 show numerical examples of the functions $T_B(v, \mu)$ and $T_B(K_{min})$. In addition, the

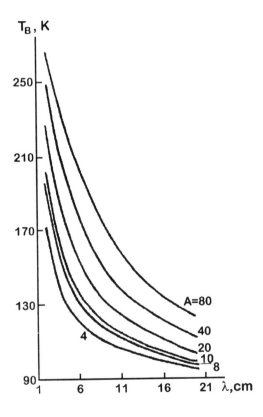

Figure 3.16. Spectral variations of the mean brightness temperature (at nadir) for isotropic rough surface. Parameters of the model: $\theta = \varphi = 0$; $\alpha = \beta = 0.01$; $\mu = v = 1$; the calculated height variances are $\sigma_\xi^2 = 0.05$–0.001 cm². The curves are plotted for several values of the spectrum constant A. There are no polarization effects.

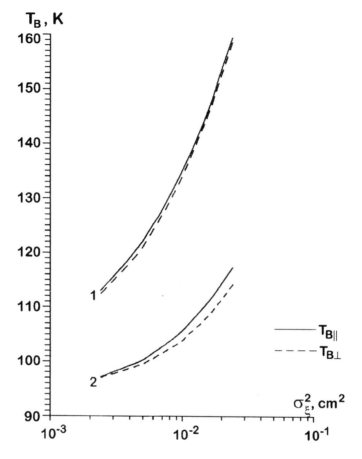

Figure 3.17. Dependencies of the brightness temperature on the high variance of the random water surface at the wavelengths: $1 - \lambda = 8$ cm; $2 - \lambda = 18$ cm. Solid line— $T_{B\parallel}(\sigma_\xi^2)$; dashed line— $T_{B\perp}(\sigma_\xi^2)$. Parameters of the model: $\theta = \varphi = 0$; $\alpha = \beta = 0.01$; $\mu = 1$; $\nu = 1.5$ (anisotropic spectrum). The effect of the polarization anisotropy $T_{B\parallel} - T_{B\perp} > 0$ has occurred.

angular functions $T_B(\theta)$ describe polarization dependencies of microwave emission (Figure 3.20).

We conclude the discussion of the analytical results with the following observation. The 'matching' mechanism is effective when a statistically good ensemble of small-scale disturbances on the ocean surface falls within an antenna footprint, and a large spatial averaging of a radiometer signal takes place. These conditions are almost always witnessed when applied to long-centimeter and decimeter wavelengths of microwave remote sensing measurements. The dynamic properties of the ocean boundary layer are known to have an appreciable influence on the conditions of generating small-scale wind-wave components and their spectral density. It is apparent that the quasi-static model used to relate the higher brightness temperature contrasts was due to the generation of short-wave capillary ripples and other overshooting-type turbulent roughness. Thus, parametrization

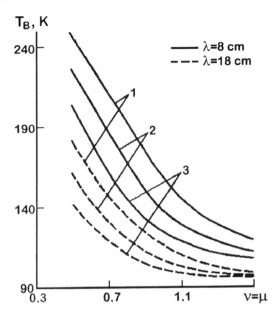

Figure 3.18. Graph of the brightness temperature vs. parameters of wavenumber spectrum $T_B(v, \mu)$. $\theta = \varphi = 0$. Solid line—$\lambda = 8$ cm; dashed line—$\lambda = 18$ cm. Parameters of the spectrum: $\mu = v$; $\alpha = \beta = 0.01$; 1—$A = 40$; 2—$A = 20$; 3—$A = 10$.

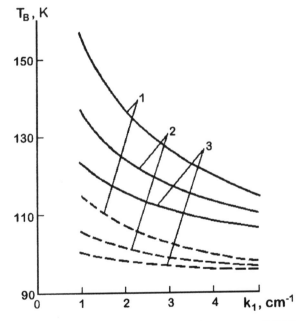

Figure 3.19. Graph of the brightness temperature vs. low-frequency cutoff $T_B(K_{min})$. $\theta = \varphi = 0$. Solid line—$\lambda = 8$ cm; dashed line—$\lambda = 18$ cm. Parameters of the spectrum: $\mu = v = 1$; $\alpha = \beta = 0.01$; 1—$A = 40$; 2—$A = 20$; 3—$A = 10$.

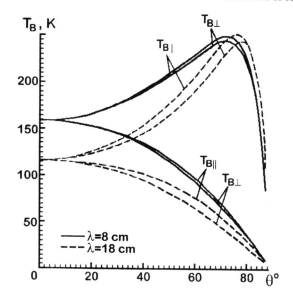

Figure 3.20. Graph of the brightness temperature vs. view angle $T_B(\theta)$. $T_{B\parallel}$ and $T_{B\perp}$—vertical and horizontal polarization. Parameters of the wavenumber spectrum: $\mu = \nu = 1$; $\alpha = \beta = 0.01$; $A = 40$.

of the high-frequency spectrum of roughness permits the modeling of different dynamical ocean surface conditions, and the estimation of corresponding effects of microwave emission. Note that principles of the macroscopic theory can also be applied to microwave ellipsometry of ocean surface roughness.

A set of aircraft ocean microwave measurements was made with the use of two radiometers at wavelengths $\lambda = 8$ and 18 cm [36]. The radiometer's system had a fluctuation sensitivity of 0.1 and 0.15 K, and a frequency band 250 and 150 MHz for $\lambda = 8$ and 18 cm wavelength channels, respectively. A unified rectangular horn antenna had a directional diagram of about 15° and 30° for $\lambda = 8$ and 18 cm channels respectively, and was pointed at the nadir view angle. The antenna spatial resolution element on the surface was about $(1/4)H$ and $(1/2)H$ for $\lambda = 8$ and 18 cm channels, respectively, where H is the flight altitude. Usually the altitude varied in the range $H = 1500–3000$ m. The radiometric receivers were of the modulated type with noisy injection. A controlled aerial photography and radar survey was carried out simultaneously with the radiometric measurements. The radiometric and auxiliary one-dimensional signals were recorded in digital real time on an onboard computer. Flights were carried out over the Pacific Ocean during 1986–1991. Methodologies of the natural radiometric experiments include the search for suitable conditions when the change of the ocean surface state was clearly observed. As a rule, long legs of 200–300 km along the general wind-wave direction, and inversely, were carried out. As a result of several experimental flights, a number of secure instances in the statistical sense were selected, which provided information about variations of the radiometric signals due to a wind-wave-generated process.

To identify the correlation between the two-channel radiometric measurements, the signals were processed using procedures of statistical and regression analysis. Figure 3.21 shows a typical example of an experimental two-channel regression graph for the brightness temperature contrasts $\Delta T_{B18}(\lambda = 18 \text{ cm})$ and $\Delta T_{B8}(\lambda = 8 \text{ cm})$, of three fixed ocean surface states. The figure shows that the regression coefficient $\rho = \Delta T_{B8}/\Delta T_{B18}$ varies as a function of the ocean boundary condition. Its values are $\rho = 2.30$, 1.76 and 1.22 for the 1–2, 3–4 and 5–6 of the Beaufort force, respectively.

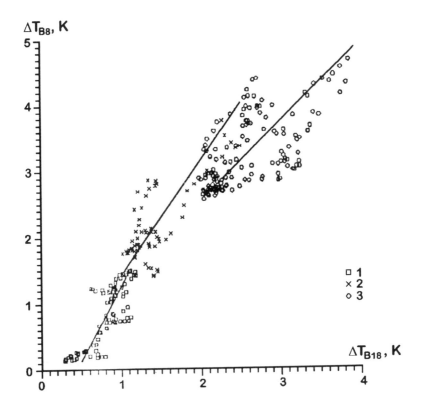

Figure 3.21. Experimental two-channel regression between the brightness temperature contrasts ΔT_{B8} and ΔT_{B18} at wavelengths $\lambda = 8$ and 18 cm (at nadir view angle). Measurements are classified by ocean surface states: '1'—1–2; '2'—3–4; '3'—5–6 of the Beaufort wind force.

The influence of atmospheric non-uniformity on the ocean emissivity at these wavelengths is small. Therefore, the variations in the brightness temperature may be explained by the change in ocean surface parameters, particularly in the change in roughness. In this case, we suggest that the main dynamical factor is a gravity–capillary wavenumber spectrum, and the regression is parametrized by wind speed.

Variations in microwave emission due to a change in the ocean surface state are described by the radiation-wind dependency of the brightness temperature contrast

$\Delta T_{\mathrm{B}}(V)$. This dependency can be modeled, first of all, by using the Mitsuyasy–Honda wavenumber spectrum for gravity–capillary surface waves:

$$S(K, \varphi) = 4.05 \cdot 10^{-3} D(U_*) K^{-3} Q(\varphi), \tag{3.30}$$

$$Q(\varphi) = \frac{1}{2\pi} [1 + 0.5 \cos 2\varphi],$$

$$D(U_*) = \left(1.247 + 2.68 \cdot 10^{-2} U_* + 6.03 \cdot 10^{-5} U_*^2 \right)^2,$$

$$U_*^2 = C_n V^2,$$

$$C_n = \begin{cases} 1.14, & 4 < V < 10 \text{ m/s}, \\ 10^{-3}(1.09 + 9.4 \cdot 10^{-2} V), & 7 < V < 17 \text{ m/s} \end{cases}$$

where U_* (cm/s) is the friction speed, C_n is the aerodynamic drag coefficient and V (m/s) is the wind speed.

Figure 3.22 shows a number of theoretical curves of the contrast $\Delta T_{\mathrm{B}}(V)$ which are optimized corresponding to experimental data. Calculations were carried out using both resonant and macroscopic models with use of the same wavenumber spectrum (3.30). Comparisons show that the macroscopic model gives the best agreement between theoretical and experimental radiation-wind dependencies $\Delta T_{\mathrm{B}}(V)$. On the other hand, the change in two-channel regression due to a variance of wind-wave conditions, which was observed in the experiment (Figure 3.21), may also be explained by the resonance model. Using a power-type wavenumber spectrum of the roughness $S(K) = AK^{-n}$, in the linear approximation, the following expression for the regression coefficient can be obtained:

$$\rho(n) = \left[\frac{k_{01}}{k_{02}} \right]^{-n+3}, \quad n \neq 1. \tag{3.31}$$

This relationship between the regression coefficient and the exponent of the power-type spectrum is simplest because the resonance functions appeared in the resonance theory,

$G\left(\dfrac{K}{k_{01}}\right)$ and $G\left(\dfrac{K}{k_{02}}\right)$, and are equal in magnitude practically for wavelengths $\lambda_1 = 8$ cm and $\lambda_2 = 18$ cm (but wave numbers are different: $k_{01} = 2\pi/\lambda_1 = 0.785$ cm^{-1}; $k_{02} = 2\pi/\lambda_2 = 0.350$ cm^{-1}). Figure 3.23 shows the diagram of the regression coefficient $\rho = \Delta T_{\mathrm{B8}}/\Delta T_{\mathrm{B18}}$ calculated for several values of the exponent n. The slope angle of the regression is easy to determine from (3.31) and is equal to $\psi = \tan^{-1}[\rho(n)]$. Comparison of all results shows that the change in the regression coefficient detected in the experiment corresponds to a change in the exponent in the range $n = 2$–3.

The detailed theoretical analysis carried out using the two electromagnetic approaches showed that both resonant narrow-band mechanism and macroscopic broad-band mechanism may contribute to the microwave emission of the rough ocean surface. The first mechanism will work predominantly in the initial stage of wind-wave generation, when

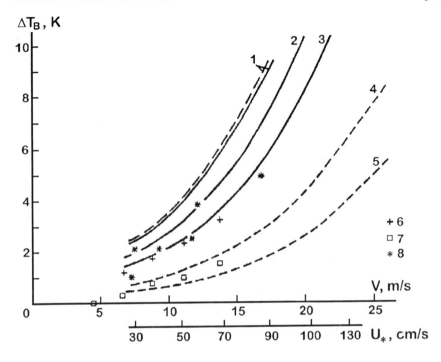

Figure 3.22. Radiation-wind dependencies of $\Delta T_B(V)$ at wavelengths $\lambda = 8$ cm (solid lines) and 18 cm (dashed lines) at nadir view angle. Calculations: 1—the resonance model. The quasi-static model for different band (43): 2—$\Delta K = 5$–100 cm^{-1} ($\Lambda < 1.25$ cm); 3—$\Delta K = 6$–100 cm^{-1} ($\Lambda < 1.05$ cm); 4—$\Delta K = 4$–100 cm^{-1} ($\Lambda < 1.57$ cm); 5—$\Delta K = 5$–100 cm^{-1}. Experimental data: 6 and 8—$\lambda = 8$ cm; 7—$\lambda = 18$ cm.

the different regular (periodic) small-scale wave structures are formed on the ocean surface (wind speed is $V < 5$ m/s approximately). The second mechanism will be more efficient at higher wind speeds, $V > 7$–10 m/s, when wave-breaking processes are started, the wave's structures become more chaotic, like a random rough surface. The choice of a preferred model depends on the purpose of the remote sensing experiment, and knowledge of the ocean boundary layer conditions.

3.3 MICROWAVE EMISSION OF FOAM AND WHITECAPS

History

The first attempts were to model foam coverage by a two-phase air–water dielectric mixture, or multi-layer structure of water and air film, by a transitional dielectric layer, and other similar models [37, 38]. Also there are a number of simple numerical approximations of the emissivity of foam coverage, depending on the wavelength and view angle (one of the first by Stogryn [39]). But all these models and numerical approximations have sometimes failed to provide sufficient precision for the analysis of experimental multi-frequency microwave data. The reason is that such approaches clearly ignore the

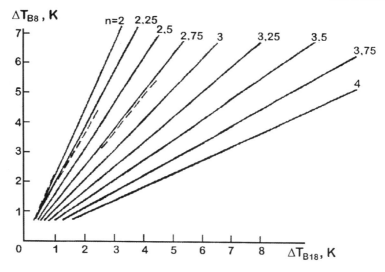

Figure 3.23. Theoretical graphs of the regression coefficient between brightness temperature contrasts ΔT_{B8} and ΔT_{B18} at wavelengths $\lambda = 8$ and 18 cm (at nadir). Solid lines—resonance model for the spectrum $S(K) = AK^{-n}$ (exponents vary in the range $n = 2$–4; $A = 0.005$–0.05). Dashed lines—experimental regression Figure 3.19.

specific nature of the dispersed oceanic medium as a system of the foam's particles interacting with the electromagnetic field at the microwave frequencies. However, the theoretical problem of scattering and emission by a densely packed system, such as that of foam or whitecaps, is very complicated, and must be solved using certain physical approximations. So, the ocean microwave remote sensing data obtained at storm conditions require the application of a more adequate model to explain, in particular, the spectral dependencies of the brightness temperature on the wind speed. It is important to note that the majority of foam coverage models used do not take an ocean's two-phase structures and its physical properties into account. For example, the laboratory microwave measurements of bubble populations and foam [40, 41] showed that there is a dependence of the reflection and emissivity on the size of the bubbles. This effect cannot be explained by using only the phenomenological models.

Experimental data

Foam and whitecaps belong to the class of colloidal systems that include two phases: gas and liquid. The physical state of foam is defined by its stability and by its inner dispersed structure. Therefore all colloids are considered heterogeneous systems with a large are of air–water interface; the foam is basically unstable. Foam that exists for a few seconds may be considered unstable, but foam that exists for some minutes or hours may be considered stable.

The proposed classification of the structures of foam and whitecaps includes:

- Mono- or polydispersed system of ideally spherical particles (gaseous bubbles), chaotically distributed in the liquid medium.
- Continuous structure of close-packed spherical bubbles.
- Cellular system of close-packed bubbles with irregular polyhedral shapes.
- Dry foams consisting of thin liquid films which are formed in polyhedral cells.

There are two categories of foam coverage in the ocean: dynamical foam (its lifetime is less than 1 min) and stable foam (its lifetime is more than 1 min). The following terminology is usually used for describing the physical conditions of the ocean foam: 'winter water', 'whitecaps' (unstable), and 'thin foam', 'foam streaks' (stable). These categories are conventional, and are based on marine observations. Special detailed investigations using the method of macrophotography show that natural stable thin foam on the sea surface represents a concentrated gas emulsion of close-packed bubbles, or 'emulsion monolayer', located on the water surface. Possible bubble diameters into the monolayer are $d = 0.01-0.5$ cm.

Detailed multifrequency microwave investigations (active and passive) of a different type of foam structure were conducted in the laboratory [42–44]. One-channel radiometers at wavelengths $\lambda = 0.26, 0.86, 2.08, 8$, and 18 cm with sensitivity of 0.1–0.2 K were used. The goal of the laboratory experiments was to measure variations in the brightness temperature induced by change in the foam structure. For this, a thick (about 1–2 cm thickness) layer of chemical foam with polyhedral cells was created on the smooth water surface. The thick layer collapsed gradually and after some time it was transformed into a stable emulsion monolayer of bubbles.

The dynamics of a foam layer were registered by radiometers continuously. Some experimental data are shown in Figures 3.24, 3.25 and 3.26. The pictures of the foam structures are shown in Figure 3.27.

During laboratory microwave experiments the following features were observed:

- Spectra of microwave emission of foam structures and foam-free water surface differed essentially. Intensity of microwave emission of foam coverage depends on its dispersed structure and thickness of the foam layer.
- At the wavelengths of $\lambda = 0.2-8$ cm, the microwave emission effect dominates owing to the bubble's monolayer (its thickness is about 0.1 cm) on the water surface.
- The wavelength range of $\lambda = 0.8-8$ cm is most sensitive to the dynamics of the foam layer and its structure transformation.
- In the wavelength range of $\lambda = 0.2-2$ cm, the thick foam layer (1–2 cm thickness) is like an 'absolute black body', i.e. the emissivity is about 1.
- At the wavelengths $\lambda = 8$ and 8 cm, variations in microwave emission were observed only in the case of a thick-enough foam layer (more than 2 cm thickness). The microwave effect from the bubble's monolayer at these wavelengths was not observed.
- At the millimeter wavelengths $\lambda = 0.26$ and 0.8 cm the emissivity is independent of polarization. Angle distribution of the emissivity is close to the diffusion law.

The scattering and absorption of microwaves by the foam structures were also investigated in the laboratory by using the method of active bistatic location [43]. The measurements of the scattering indicatrix from the foam layer located on the water

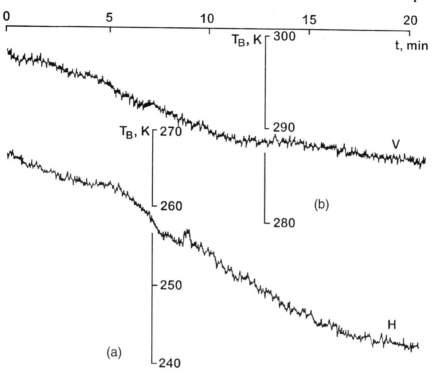

Figure 3.24. Temporal radiometric signal from collapsed foam layer at the wavelength $\lambda = 0.86$ cm. Angle of view is $\theta = 35°$. Polarization: vertical (V) and horizontal (H). Time of measurements: $\tau = 0$—polyhedral-cells foam (a); $\tau = 20$ min—stable bubble's monolayer (b).

surface were made at the frequencies 9.8, 36.2, and 69.9 GHz. It was found that an indicatrix is mainly determined by the foam structure. Figure 3.28 shows some results of the bistatic measurements (scattering indicatrix).

At the frequencies 36.2 and 69.9 GHz (corresponding wavelengths are $\lambda = 0.83$ and 0.43 cm) the reflected signal from the thick (1–15 cm) layer of polyhedral-cell foam was $\leqslant -25$ dB. This value gives the estimation of the power reflection coefficient as $\leqslant 0.005$. At the frequency 9.8 GHz ($\lambda = 3$ cm) the same layer of foam gives only a specularly reflected component. However, the scattering indicatrix of the emulsion monolayer has several resonant maximums. The relative levels of the scattered signal at the frequencies 36.2 and 69.6 GHz are $\leqslant -12$ dB. In this case the power reflection coefficient was ~ 0.03–0.06, and at the frequency 9.8 GHz the power reflection coefficient was ~ 0.6.

Active microwave measurements show that foam structures are singular 'black bodies'. This is associated with a high absorption of microwaves by bubbles and liquid cells of foam. Resonant effects of the scattering from a thin (~ 0.1 cm) emulsion mono-layer are typical for a diffraction grating with a period of about the wavelength $d \sim \lambda$. In fact, the emulsion monolayer represents a regular two-dimensional grating whose effective nodes are the bubbles floating on the water surface. We can indicate the analogy

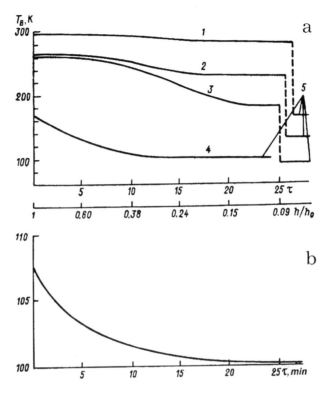

Figure 3.25. Temporal change of brightness temperature during decay of foam layer on the water surface. View angle: $\theta = 35°$. Data of laboratory multi-channel radiometric measurements: (a) 1—$\lambda = 0.86$ cm, vertical polarization; 2—$\lambda = 0.86$ cm, horizontal polarization; 3—$\lambda = 2.08$ cm, horizontal polarization; 4—$\lambda = 9$ cm, horizontal polarization; 5—level of free-foam flat water surface; (b) $\lambda = 18$ cm, horizontal polarization. The foam thickness changes according to the exponential law: $h = h_0 \, e^{-k_r \tau}$.

between the observed resonant phenomena and the diffraction of X-rays at crystal lattices or polyatomic molecules of liquids.

Theoretical models

A theoretical explanation of the microwave effects was carried out by using a new model of close-packed two-layer concentric spheres [44]. It was found that such particles in the microwave range possess specific radiation characteristics. Special numerical analysis of the scattering and absorption cross-sections for the hollow spherical particle or water shell (bubble) in the microwave range was made by using the Mie theory. The calculation shows that in the wavelength range of $\lambda = 0.2$–0.8 cm, large-sized bubbles with a diameter of $d = 0.1$–0.2 cm are resonant objects like a multipole with strong absorption and scattering. But in the wavelength range of $\lambda = 2$–8 cm and higher, all bubbles with a diameter of $d < 0.2$ cm are like a dipole only. These are the Rayleigh particles. For such bubbles the absorption cross-sections essentially exceed the scattering cross-sections

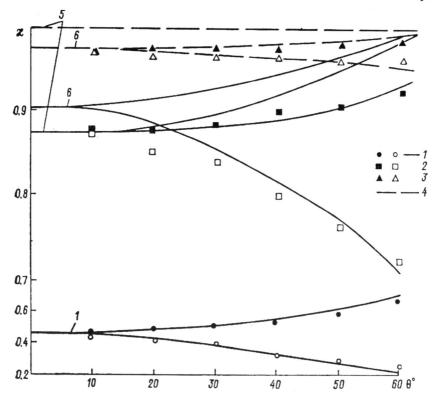

Figure 3.26. Dependencies of foam emissivity on view angle at wavelength $\lambda = 0.86$ cm. Experimental data: 1—free-foam flat water surface; 2—emulsion monolayer; 3—polyhedral-cells foam; 4, 5 and 6—calculations by using the models # 5 and 6 (Table 3.1) for emulsion monolayer (solid line) and polyhedral-cells foam (dashed line).

(Figure 3.29). This fact is taken into account in the development of the new microwave model of foam and whitecaps.

Modeling of foam/whitecaps microwave emission includes the following steps:

- Choice of input parameters (geometry and size distribution of bubbles).
- Calculation of radiation characteristics of bubbles by using the Mie theory.
- Calculation of effective permittivity of the polydispersed system of bubbles.
- Calculation of reflection coefficient from a multilayer system with the vertical distribution of the effective permittivity.
- Calculations of microwave emission spectrum.

The main parameter of the suggested microwave model is an effective permittivity of the foam/whitecap media. To define this parameter, the well-known, from molecular optics, Lorenz–Lorentz formula and Hulst equation [45] were modified and were used for the polydispersed system of the bubbles. The first formula (3.32) takes into account dipole interaction of bubbles in a close-packed dispersed system (the quasi-static approximation). The second formula (3.33) describes the contribution of the multipole moment

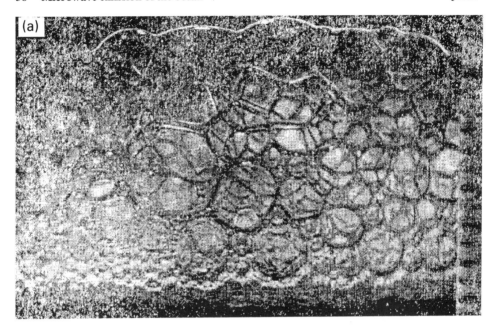

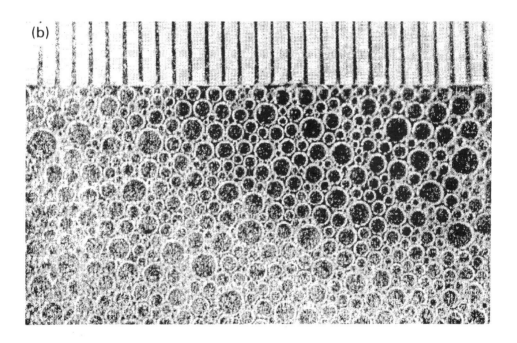

Figure 3.27. Pictures of foam structures: polyhedral-cells foam (a) and emulsion monolayer of bubbles (b). The graduation line is equal to 0.1 cm.

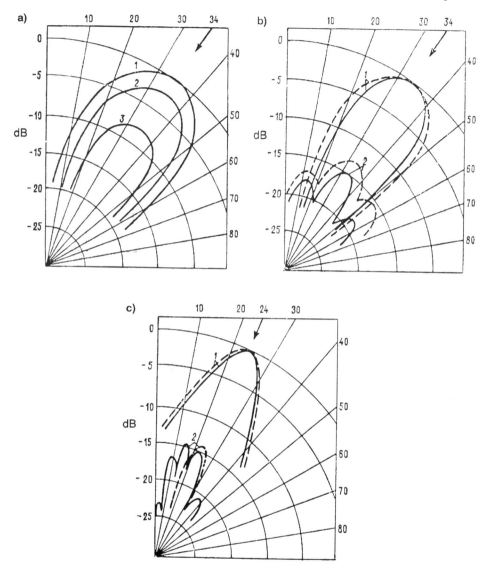

Figure 3.28. Experimental scattering indicatrix of foam structures. 1—aluminum plate; 2—emulsion monolayer; 3—polyhedral-cells foam; solid line—vertical polarization; dashed line—horizontal polarization. (a) $\lambda = 3$ cm (9.8 GHz), $\theta = 34°$; (b) $\lambda = 0.83$ cm (36.2 GHz), $\theta = 34°$; (c) $\lambda = 0.43$ cm (69.9 GHz); $\theta = 24°$.

of bubbles into effective permittivity of the system (the approximation of the electrodynamically 'dilute' medium). These approaches are given by the following formulas.

Modification of the Lorenz–Lorentz equation gives

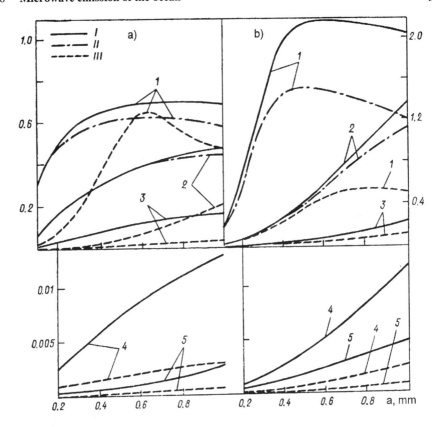

Figure 3.29. Effective factors of extinction (I), absorption (II), and scattering asymmetry (III) depending on outer radius of the bubble. Thickness of water shell: (a) $\delta = 0.001$ cm; (b) $\delta = 0.005$ cm. Electromagnetic wavelength is changed: 1—$\lambda = 0.26$ cm; 2—$\lambda = 0.86$ cm; 3—$\lambda = 2.08$ cm; 4—$\lambda = 8$ cm; 5—$\lambda = 18$ cm. (Calculations were made by L. Dombrovskiy using the Mie formulas.)

$$\varepsilon_{N\alpha} = \frac{1 + \frac{8}{3}\pi\overline{N\alpha}}{1 - \frac{4}{3}\pi\overline{N\alpha}}, \qquad \overline{N\alpha} = \frac{k\int \alpha(a)f(a)\,\mathrm{d}a}{\frac{4}{3}\pi\int a^3 f(a)\,\mathrm{d}a}, \tag{3.32}$$

$$\alpha = a^3 \frac{(\varepsilon_0 - 1)(2\varepsilon_0 + 1)\,(1 - q^3)}{(\varepsilon_0 + 2)(2\varepsilon_0 + 1)(1 - q^3) + 9\varepsilon_0 q^3}.$$

Modification of the Hulst equation gives

$$\varepsilon_{NS} = 1 + \mathrm{i}4\pi\left(\frac{2\pi}{\lambda}\right)^{-3}\overline{NS_0}, \qquad \overline{NS_0} = \frac{k\int S_0(a)f(a)\,\mathrm{d}a}{\frac{4}{3}\pi\int a^3 f(a)\,\mathrm{d}a}, \tag{3.33}$$

$$S_0 = \sum_{n=1}^{\infty} \frac{2n+1}{2}(A_n + B_n),$$

where:

$\varepsilon_{N\alpha}$ and ε_{NS} are complex effective permittivities of the polydispersed system;
$f(a)$ is the normalized size-distribution function;
a is the external radius of a single bubble;
δ is the thickness of the shell;
N is the volume concentration of the bubbles;
k is the packing coefficient of the bubbles;
ε_0 is the complex permittivity of the shell medium (usually salt water);
α is the complex polarizability of a single bubble;
S_0 is the complex amplitude of the scattering 'forward' by a single bubble;
$q = 1 - \delta/a$ is the bubble's 'filling' factor;
A_n and B_n are the complex Mie coefficients for a hollow spherical particle.

Figure 3.30 shows the spectral dependencies of the real and imaginary parts of the effective complex permittivity $\varepsilon_{N\alpha}(\lambda)$ for different parameters of foam bubbles according to the Lorenz–Lorentz approximation (3.32). The shaded field in Figure 3.30 is limited by two curves which correspond to calculations using the traditional electrodynamic model of foam as a two-phase statistical mixture of water and air. It is seen from Figure 3.30 that $\varepsilon_{N\alpha}(\lambda)$ changes over a wide range. The imaginary part of $\varepsilon_{N\alpha}$ has a maximum in the range $\lambda = 0.8$–2 cm, and exceeds the corresponding quantity ε_{NS}.

Figure 3.31 shows the complex permittivity of foam $\varepsilon_{NS}(\lambda)$ according to the *Hulst* approximation (3.33). The complex amplitude of the scattering 'forward' $S_0(\lambda)$ was calculated by using the Mie formulas for hollow spherical particles (bubbles). It follows from Figure 3.31 that a monodispersed system of bubbles has a maximum absorption in the range of wavelengths $\lambda = 0.8$–3 cm. The maximum of the imaginary part $\varepsilon_{NS}(\lambda)$ is due to the multipole moments of water bubbles at microwave frequencies. It is interesting to note that the first resonant Mie effects which are associated with the oscillations of the amplitude $S_0(\lambda)$ occur for the bubble's system at condition $a \sim \lambda/4$.

Taking into account the dipole or multi-pole moment of the bubbles therefore leads to an increase in both the real and the imaginary parts of the foam effective permittivity. In other words, full electromagnetic losses in the foam caused by the diffraction mechanism increase appreciably. It is also characteristic that these conditions are realized in a sufficiently wide microwave range. This effect cannot be achieved by using simple models of a dielectric mixture which exclude from consideration effects associated with the specific properties of bubbles at microwave frequencies.

In the simplest case of a flat surface, emissivity of the foam–water system is equal to $\kappa_f(\lambda) = 1 - \left| r_f(\lambda) \right|^2$, where $r_f(\lambda)$ is the spectral reflection coefficient of a multi-layer medium with the arbitrary complex function $\varepsilon_{N\alpha}(z)$ or $\varepsilon_{NS}(z)$, and where z is the depth of an elementary layer in a multi-layer system. The distribution of $\varepsilon_{N\alpha}(z)$ or $\varepsilon_{NS}(z)$ depends on the vertical distribution of the phase components (water and air) or vertical distribution of the bubble's size. Here we illustrate only two working electromagnetic models that correspond to the dispersed systems of the polyhedral and emulsion types (Figure 3.27). They are like natural structures and are called 'foam streaks' (stable sea foam) and 'whitecaps' (unstable sea foam). These models are schematically shown in Figure 3.33.

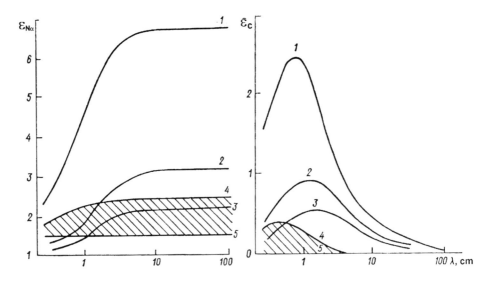

Figure 3.30. Spectral dependencies of effective permittivity of foam. Modeling by using Lorenz–Lorentz approximation (3.32) with parameters: temperature $t = 20°C$; salinity $s = 0$ ppt; thickness of bubble shell: 1—$\delta = 0.01$ cm; 2—$\delta = 0.005$ cm; 3—$\delta = 0.0001$ cm; 4 and 5—calculation by using dielectric mixing theory [46] for volume concentration of water in the air–water dielectric mixture $C = 0.2$ and 0.01 respectively.

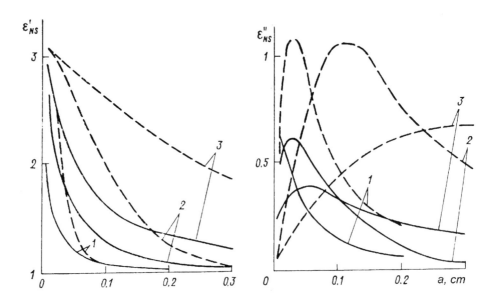

Figure 3.31. Dependencies of effective permittivity on external radius of foam's bubble. Calculation by using the Hulst approximation (3.33) at the wavelengths: 1—$\lambda = 0.26$ cm; 2—$\lambda = 0.86$ cm; 3—$\lambda = 2.08$ cm. Solid line—$\delta = 0.001$ cm; dashed line—$\delta = 0.005$ cm ($t = 20°C$; salinity $s = 0$ ppt).

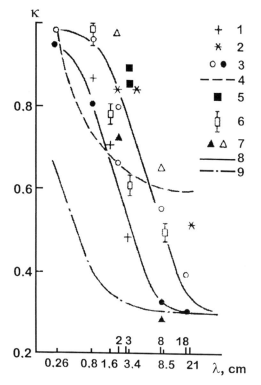

Figure 3.32. Emissivity of the foam structures. Experimental data: 1—airborne; 2—onshore; 3 and 5—laboratory; 4—numerical approximation [39]; 6 and 7—shipborne; 8—two models (Figure 3.33); 9—smooth foam-free water surface.

The first model is a two-layer structure with a smeared lower boundary. A thin uniform dielectric layer represents a mono- or polydispersed system of bubbles located off the smooth water surface. The second model envisages a vertical distribution of bubbles with different sizes. A thick non-uniform dielectric layer of foam is characterized by a variation in the bubble's radius from a fraction of a millimeter near the water surface, to a centimeter at the foam–atmosphere interface. As the height of the layer increases, the bubbles lose their spherical shape, and turn into polyhedral cells that are kept together by liquid films. By choosing the dependence of the bubble's radius $a(z)$ to be linear and using the Hulst equation, we obtain a continuous vertical distribution of the effective permittivity $\varepsilon_{NS}(z)$.

The comparative list of possible microwave models, which were applied for the analysis of the laboratory and natural microwave data, is shown in Table 3.1. The last model (# 6) describes macroscopic properties of foam/whitecaps. Different parameters of the model such as the thickness of the dispersed layer, geometry, and size-distribution of bubbles are used for separation of foam and whitecaps microwave signatures. Experimental data and results of modeling are shown in Figure 3.32.

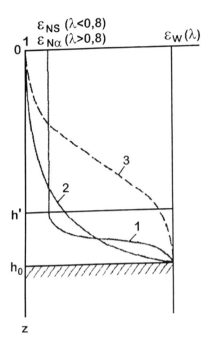

Figure 3.33. Vertical distribution of the effective permittivity for different types of foam structure. Examples of the models: 1—emulsion monolayer of close-packed bubbles on the water surface (foam streaks); 2—polyhedral-cell's structure (whitecaps); 3—transitional dielectric layer with smoothly changed parameters under the hyperbolic tangent law.

As we have already noted, models of the uniform dielectric layer type with parameters corresponding to the heterogeneous mixture of water and air do not give a quantitative agreement with experiment. This is also true for models of a discrete-layered structure of water and air films, and for smooth transition-layer models. In the model of a discrete scattering medium, one takes into account the diffraction properties of bubbles at microwave frequencies. This makes it possible to obtain a better agreement with experiment. But 'optical' models of foam are based on the solution of the radiation transfer equation, i.e. they assume a complete incoherence of the scattering and absorption events. Application of the radiation transfer equation for the description of foam/whitecaps as a closely packed system of scattering particles requires an additional investigation. The 'optical' model also does not qualitatively describe the polarization characteristics of foam that are experimentally observed (Figure 3.26).

Among the first shipborne microwave measurements of oceanic foam/whitecaps made were by the use of a two-frequency radiometer operating at the wavelengths $\lambda = 2$ and 8 cm with fluctuation sensitivity of ~ 0.05 K [47]. Shown in Figure 3.34 is a fragment of a synchronous radiometric recording in probing the sea surface while the ship was in motion. The amplitude of the radiometric signal depends upon the relative area of the foam/whitecap-covered patches within the footprint of the antenna. The experiments made it possible to evaluate the microwave effects of foam/whitecaps. In the case of

Table 3.1. Electrodynamic models of the foam/whitecaps

		Agreement with experiment	
#	Model name	Foam (thin layer, $h = 0.1$–0.5 cm)	Whitecaps (thick layer $h = 2$–5 cm)
1.	A layer of the air–water dielectric mixture	No	No
2.	A multi-layer structure from uniform films of the water and air	No	No
3.	A non-uniform dielectric layer of air–water mixtures with the vertical distribution of phase components	No	Yes
4.	Transitional dielectric layer with smoothly changed parameters under the hyperbolic tangent law	No	Yes
5.	Discrete scattering media consisting of spherical bubbles with specific diffraction parameters. Radiation transfer equation is used	Yes	No
6.	Non-uniform multi-layer medium with effective dielectric parameters depending on diffraction properties and size distribution of the bubbles. Quasi-static approaches used	Yes	Yes

whitecaps (wave crest breaking) there were maximum increments of the brightness temperature $\Delta T_B = 120$ and 80 K; in the case of foam streaks, the values were $\Delta T_B = 70$ and 40 K, respectively, in the $\lambda = 2$ and 8 cm channels. These values were obtained when the foam/whitecaps' structures practically filled the radiometer antenna beam. The spatial resolution was 0.5 and 15 m for $\lambda = 2$ and 8 cm (view angle was 55° from nadir; the radiometer was located in the ship 7 m above sea level). An estimation of the foam/whitecaps' emissivity was made from the ship's radiometric data. A good correlation was achieved with the data from microwave laboratory experiments, and macroscopic models which were described in this section.

We may conclude that the suggested macroscopic models of foam/whitecaps with the introduction of the dipole (multipole) moment of closely packed bubbles' system clearly reflect most completely the microwave properties of oceanic dispersed media. In the dipole approximation, the cooperative effects and interactions between particles play an important role. They associate with the dense packing of the two-layer particles. This approximation can be used to advantage in the analysis of multi-frequency experimental data in the centimeter and decimeter ranges of wavelengths. In the millimeter range, the

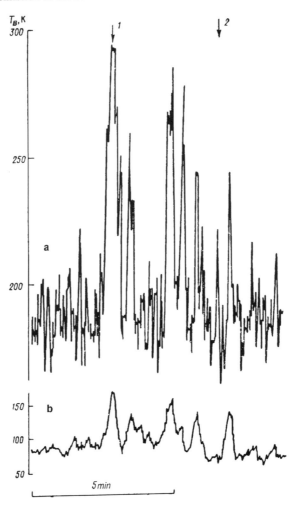

Figure 3.34. Fragment of two-channel microwave radiometer recording of foam-covered sea surface. Angle of view $\theta = 55°$. Variations of brightness temperature at the wavelengths: (a) $\lambda = 2$ cm (vertical polarization); (b) $\lambda = 8$ cm (horizontal polarization). Marks: 1—whitecap; 2—foam streak.

multipole approximation gives more adequate results. However in this case, there is a general electromagnetic problem associated with an adequate description of densely packed systems like foam/whitecaps. The development of the macroscopic models can be made by using a representation of the foam structure as fractal (multi-fractal) clusters of the bubbles.

3.4 EMISSIVITY OF SPRAY

The structure of ocean spray has been investigated in nature with sufficient accuracy [48, 49]. It has been found, in particular, that the size distribution of the ocean spray follows the power law $p \sim r^{-n}$, where r is the radius of a droplet. The exponent n changes from 2 to 8, depending on wind conditions. The range of a droplet's diameter is quite wide: $d = 10^{-4}–10^{-2}$ cm. The height of a dense spray layer over the ocean surface is about 10–40 cm and depends significantly on the droplets' generating mechanism. The mass concentration of the water in the ocean spray near the surface is about $\varpi = 10^{-4}–10^{-1}$ g/cm³. Dense layers of spray are located mostly around the breaking crest of wind waves (whitecaps). The vertical distributions of size and concentration of spray are highly non-uniform. Clouds of small-sized droplets and aerosol are formed over the foam-free water surface. But large-sized droplets cover vast expanses of the areas of foam and whitecaps.

To describe the size distribution of the spray, the universal Γ-function can be introduced:

$$p(r) = \frac{A^{B+1}}{\Gamma(B+1)} r^B \, e^{-Ar}, \tag{3.34}$$

where A and B are the variant parameters. The 'tails' of the distribution are very sensitive to changes in wind speed and spray-generated conditions.

Modeling of the microwave characteristics of spray is based on the solution of the radiation transfer equation for a discrete scattering and absorbing medium which contains the spherical water droplets made up of different sizes. In the simple case of hemispherical emissivity of the dispersed system, the analytic solution of the radiation transfer equation can be used [50].

For a polydispersed system of water droplets, volume factors of the absorption $\overline{Q}_a$, scattering $\overline{Q}_s$, and attenuation $\overline{Q}_t$ are introduced:

$$\{\overline{Q}_a, \overline{Q}_s, \overline{Q}_t\} = \frac{3}{4} \frac{\varpi}{\rho} \{Q_a, Q_s, Q_t\} r^2 p(r) \, dr \Big/ \int r^3 p(r) \, dr, \tag{3.35}$$

where Q_a and Q_s are the effective factors of absorption and scattering; $Q_t = Q_a + Q_s$ is the effective factor of attenuation for the salt water droplet of radius r; ϖ is the mass concentration of the water; ρ is the density of the salt water.

In the microwave range, factors of Q_a, Q_s and Q_t are calculated by using the Mie theory for the spherical particle (water droplet) with diffracted parameter $x = 2\pi r/\lambda$. Note that these factors are the function of the complex permittivity of the water and depend on its temperature and salinity. Some calculations for spray are shown in Figure 3.35.

The main electromagnetic properties of water droplets are as follows:

- In the microwave range $\lambda > 0.6$ cm, small sized droplets (of radius $r < 0.05$ cm) are particles that obey the Rayleigh law of scattering.
- The resonant region of scattering and absorption for large-sized water droplets (of radius $0.05 < r < 0.2$ cm) is manifested in the wavelength range of $\lambda = 0.3–5$ cm.
- In the wavelength range of $\lambda = 0.2–8$ cm the radiation properties of the droplets depend on the water temperature.

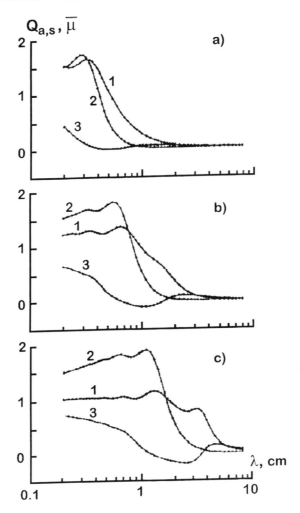

Figure 3.35. The effective factors of absorption (1), scattering (2), and asymmetry of scattering (3) for spherical water droplets of radius: (a) $r = 0.05$ cm; (b) $r = 0.1$ cm; (c) $r = 0.2$ cm.

The emissivity of spray is determined not only by diffracted characteristics of individual water droplets, but also by the mass concentration of water in the spray layer. For example, let us consider the standard two-layer model in which the flat layer of spray is located on a smooth (foam free) water surface. Spray is represented by a discrete polydispersed system of water droplets with complex dielectric constant. The corresponding solution of the scalar radiation transfer equation gives the following formulas for the spectral hemispherical emissivity [51]:

$$\bar{\kappa} = 4\sum_{i=1}^{4}(-1)^i \Delta_i E_i \Big/ \sum_{i=1}^{4}(-1)^i \Delta_i a_i, \qquad E_i = \exp(\Lambda_i \tau_0); \tag{3.36}$$

$$\Lambda_{1,2} = \sqrt{6\left(1 + \alpha \pm \sqrt{1 + \alpha + \alpha^2}\right)}; \qquad \Lambda_{3,4} = -\Lambda_{1,2},$$

$$\Delta_{1,2} = \begin{vmatrix} b_{2,1} & b_3 & b_4 \\ c_{2,1} & c_3 & c_4 \\ d_{2,1} & d_3 & d_4 \end{vmatrix}, \qquad\qquad \Delta_{3,4} = \begin{vmatrix} b_1 & b_2 & b_{3,4} \\ c_1 & c_2 & c_{3,4} \\ d_1 & d_2 & d_{3,4} \end{vmatrix},$$

$$a_i = \left[\frac{\Lambda_i^2}{6\alpha} + \frac{\Lambda_i}{\alpha} + 1\right]E_i, \qquad\qquad b_i = \left[\frac{\Lambda_i^3}{6\alpha} + \frac{\Lambda_i^2}{2\alpha} - 2\Lambda_i - 3\right]E_i,$$

$$c_i = -\frac{\Lambda_i^2}{6\alpha} + \bar{\kappa}_0\frac{\Lambda_i}{\alpha} - 1, \qquad\qquad d_i = \bar{\kappa}_0\frac{\Lambda_i^3}{6\alpha} - \frac{\Lambda_i^2}{2\alpha} - 2\Lambda_i + 3,$$

$$\bar{\kappa}_0 = \frac{\kappa_0}{2 - \kappa_0}.$$

Here, $\alpha = \Sigma_a/\Sigma_{tr}$ is the ratio of the absorption coefficient to the transport attenuation coefficient; $\tau_0 = \Sigma_{tr}h_s$ is the optical thickness of the dispersed layer, and $\bar{\kappa}_0$ is the hemispherical emissivity of the underlying medium (water, foam, etc.). Let us consider some results of the modeling by using (3.34)–(3.36).

Figure 3.36 shows the spectral dependencies of emissivity $\kappa(\lambda)$ of the spray–water system for the case of a monodispersed layer of droplets with a different surface mass concentration $\varpi h = 0.1$ and 0.01 g/cm^2 (h is the height of the spray layer). For large-sized droplets (radius $r = 0.2$ cm) the curves $\kappa(\lambda)$ are reminiscent of the spectral dependencies $Q_a(\lambda)$. This correspondence is not accidental, since the scattering in the Rayleigh region is weak compared with the absorption.

Figure 3.37 shows the analogous dependencies for the case of a polydispersed system of droplets with different size-distribution parameters (3.34). The important variable parameter here is the surface mass concentration ϖh.

The main features of the emissivity of the spray–water system are as follows:

- Owing to spray, the emissivity increases steeply in the wavelength range $\lambda = 0.2$–5 cm.
- Microwave signatures of spray reflect the fluctuations (variability) of the size and mass concentration of water. The higher the mass concentration of the water, the higher the emissivity of the system.
- In the case of the monodispersed layer, resonant effects are pronounced, while for the polydispersed layer they are smoothed out.
- The emissivity of the spray is determined mainly by large-sized droplets, i.e. by the 'tails' of the size distributions (3.34).

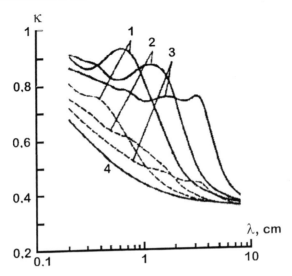

Figure 3.36. Spectrum of emissivity for monodispersed spray with different droplet radii: 1—$r = 0.05$ cm; 2—$r = 0.1$ cm; 3—$r = 0.2$ cm; 4—smooth water surface. Solid line: $\varpi h = 0.1$ g/cm^2. Dashed line: $\varpi h = 0.01$ g/cm^2.

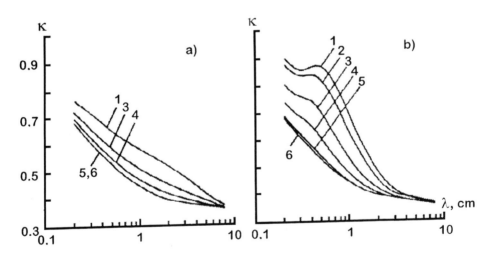

Figure 3.37. Spectrum of the emissivity for polydispersed spray with different surface mass concentrations: 1—$\varpi h = 0.1$ g/cm^2; 2—$\varpi h = 0.08$ g/cm^2; 3—$\varpi h = 0.05$ g/cm^2; 4—$\varpi h = 0.03$ g/cm^2; 5—$\varpi h = 0.01$ g/cm^2; 6—smooth water surface. (a) Small-sized spray ($r_{max} \approx 0.01$ cm); (b) large-sized spray ($r_{max} \approx 0.1$ cm).

Detailed laboratory and shipborne remote sensing investigations of the laboratory droplet's flow and natural oceanic spray induced by breaking waves were made by using one radiometer/scatterometer at a wavelength of $\lambda = 0.8$ cm.

Figure 3.38 shows some results of microwave laboratory experiments. Characteristics of microwave propagation such as coefficients of backscattering and extinction, also brightness temperature, were measured synchronously for different parameters of the water droplet's medium. It was found that the droplets flow like discrete scattering media

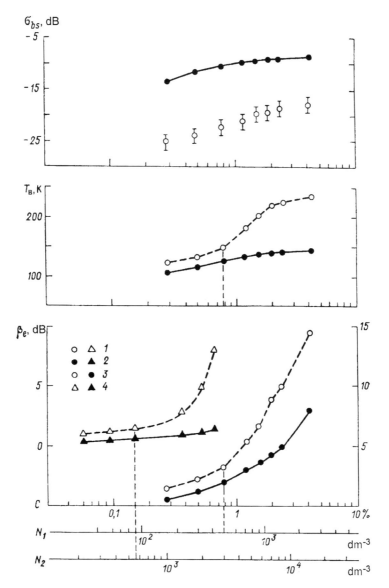

Figure 3.38. Dependencies of backscattering coefficient (σ_{bs}), brightness temperature (T_B), and extinction coefficient (β_e) on piece concentration ($N_{1,2}$) of a droplet's flow. 1—laboratory measurements at the wavelength $\lambda = 0.8$ cm; 2—standard radiation transfer theory; 3 and 4—droplet's flow with different dispersed characteristics.

at a low concentration ($c \leqslant 0.1\%$) of droplets. In this case the microwave properties of the media (thermal emission and extinction) can be described by a standard radiation transfer theory with the use of the Mie resonant formulas. The theory, however, cannot explain the backscattering features of the media because a coherent part of the scattering is not taken into account. In the case of high concentration ($c \leqslant 4.5\%$) of droplet's media, the microwave properties are like a continuous random medium with a strong fluctuation of the permittivity. This medium is an analog of natural oceanic spray. Figure 3.39 shows the spectrum of microwave signal intensity which was measured at the backscattering and direct propagation regimes. The spectrum has an exponent of '$-5/3$' and corresponds to the turbulent medium.

Processes of wave breaking and formations of foam–spray in nature were investigated from the ship by using the same microwave technique [52]. The dynamics of wave breaking were measured by a radiometer–scatterometer at wavelength $\lambda = 0.8$ cm. Figures 3.40 and 3.41 illustrate the fragment of temporal synchronous radiometer–scatterometer records. These data show how the signals from foam and spray are separated at the detection of the wave-breaking process. In fact, that backscattering cross-section σ_{bs} and brightness temperature T_B are changed with anti-phase. The value of σ_{bs} increases faster than the value of T_B. A maximum of backscattering and a minimum of microwave emission are observed at the point when the breaker has occurred. Consequently, the backscattering signal is sensitive to spray injection, and the microwave signal is sensitive to foam/whitecaps. Disappearance of the spray and appearance of the stable foam coverage cause a decrease in the backscattering and an increase in the microwave emission of the ocean surface. The observed effect can be explained by the use of a combined foam–spray microwave model (paragraph # 6).

3.5 CONTRIBUTION OF BUBBLE POPULATIONS

In view of microwave remote sensing, the bubble populations can be considered as an intermediate type of structure between a surface foam monolayer and an underwater bubble distribution. Bubble populations are located at the air–water surface as single clusters or a collection of such clusters. Sometimes an interference picture from films of bubbles is observed. Mathematically this surface configuration can be represented by an ensemble of small- or large-sized hemispherical shells floating on the water surface. At microwave frequencies the influence of the dielectric subsurface layer and diffraction properties of bubbles should be considered simultaneously. In this case the electrodynamic problem becomes significantly complicated. In a simple case the statistical model of random oriented water films and approximation of geometrical optics can be used. Also the model of a single dipole (multipole) located over the dielectric surface can be applied. Microwave signatures of these objects can reflect non-uniformity of the surface density of a bubble's cluster. This cluster can act like bubble's fractal or multi-fractal object. So, the emissivity and scattering from these objects may have features associated with resonant effects.

Another two-phase system is formed owing to generation of underwater bubble populations. Mechanisms of aeration of the ocean water and appearance of gaseous

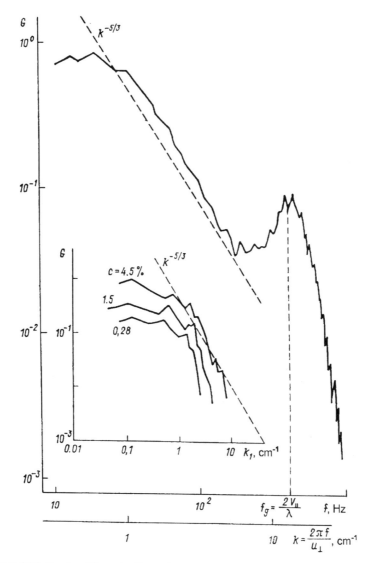

Figure 3.39. Spectra of intensity fluctuations of microwave backscattering (a) and 'forward' propagation (b) for a droplet's flow with different values of the volume concentration $c = 4.5\%$ and $c = 0.28$–4.5%, respectively. Laboratory measurements at wavelength $\lambda = 0.8$ cm. The velocity component of droplets in the antenna direction in backscattering mode was $V_{\parallel} = 1.7$ m/s.

bubbles in the ocean subsurface (at a depth < 1 m) may be connected with the following processes:

- By mechanical mixing of air and water due to wave breaking;
- By migration of gaseous bubble flow from the deep water to the surface due to natural chemical reactions;
- Air-caused cavitation due to the rotation of blades of a ship's propeller (hydrodynamic cavitation);

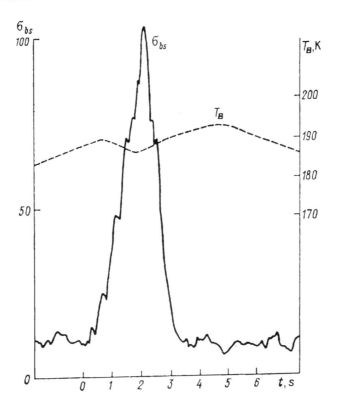

Figure 3.40. Temporal synchronized scatterometer and radiometer records of wave-breaking process (unit ocean wave). Shipborne measurements at the wavelength $\lambda = 0.8$ cm. σ_{bs}—backscattering coefficient; T_B—brightness temperature.

- Propagation of intense sound waves in oceanic water (acoustic cavitation);
- Cavitation currents induced as an effect of strong flow over a moving body.

Experimental results of investigations of these processes in the open ocean have not been published except for some experimental data about the results of acoustic sounding underwater bubbles and measurements of the ambient noise [53, 54], as well as some measurements of bubble distributions on a deep water *in situ* (for example, Mulhearn [55]). These investigations show that the size distribution of bubbles is governed by the power law $p \sim a^{-n}$, where n varies between 3.5 and 5.5 at the range of a bubble's radius of $a = 10^{-4}$–10^{-1} cm. The volume concentration of gaseous bubbles in the upper layer of the ocean can reach 20% or more. The process of aeration has an influence on the condition of the air–water interface, and it therefore changes the electrodynamical properties of the ocean skin-layer.

The estimations of a possible microwave effect due to aeration of the water subsurface can be made by using a simple quasi-static model of the two-phase dielectric mixture. The complex effective permittivity of the air–water matrix mixture can be represented by the common formula [56]

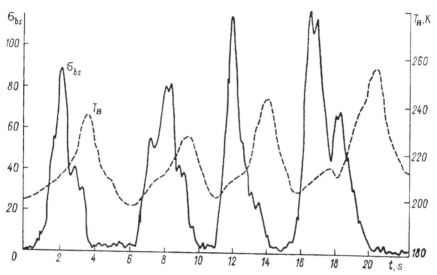

Figure 3.41. Temporal synchronized scatterometer and radiometer records of ship's wave breaking. Shipborne measurements at the wavelength $\lambda = 0.8$ cm. View angle $\theta = 80°$, vertical polarization. σ_{bs}—backscattering coefficient; T_B—brightness temperature.

$$\varepsilon_m = \varepsilon_w + c(\varepsilon_i - \varepsilon_w)\sum_{j=1}^{3}\frac{1}{1+\left(\dfrac{\varepsilon_i}{\varepsilon^*}-1\right)A_j},$$

$$(3.37)$$

where $\varepsilon_w(\lambda)$ is the complex permittivity of the water; $\varepsilon_i = 1$ is the permittivity of air (or any gas); c is the volume concentration of bubbles in the water. Formula (3.37) takes the geometrical form of bubbles over the form-factor A_j into account. For example, for spheres it is $\{A_1 = A_2 = A_3 = \frac{1}{3}\}$; for needles—$\{A_1 = A_2 = \frac{1}{2}, A_3 = 0\}$; and for disks—$\{A_1 = A_2 = 0, A_3 = 1\}$. The parameter ε^* is the effective permittivity of the medium around bubbles and describes some further contribution of the close electrostatic field to the effective permittivity of the mixture. Because the value of ε^* is not known exactly, the following two approaches are used: $\varepsilon^* = \varepsilon_w$ and $\varepsilon^* = \varepsilon_m$. By substituting both values into (3.37), we obtain two boundaries for the function $\varepsilon_m(c)$. These boundaries enable us to estimate a probable region for the value ε_m.

Also the formula (3.37) may be rearranged to give the real and imaginary parts of the complex permittivity $\varepsilon_m = \varepsilon_m' + i\varepsilon_m'$:

$$\varepsilon_m' = \varepsilon_{m\infty} + \frac{\varepsilon_{m0} - \varepsilon_{m\infty}}{1+\left(\dfrac{\lambda_{ms}}{\lambda}\right)^2},$$

$$\varepsilon_m'' = \frac{\varepsilon_{m0} - \varepsilon_{m\infty}}{1+\left(\dfrac{\lambda_{ms}}{\lambda}\right)^2}\cdot\frac{\lambda_{ms}}{\lambda}.$$

$$(3.38)$$

These relations describe the effective permittivity of a two-phase medium as a liquid dielectric with parameters of relaxation: ε_{m0}, $\varepsilon_{m\infty}$, λ_{ms} analogously to Debye's equations for dielectric constants of fresh water.

Figure 3.42 contains some diagrams $\varepsilon_m''(\varepsilon_m')$ calculated for the air-bubbles-in-water emulsion. The values of air concentration $c = 0.05$–0.1 are close to the true values. All diagrams are designed by using formula (3.38) when the wavelength λ is changed gradually from 0.1 cm to 30 cm. The value $c = 0$ corresponds to the case of air-free water.

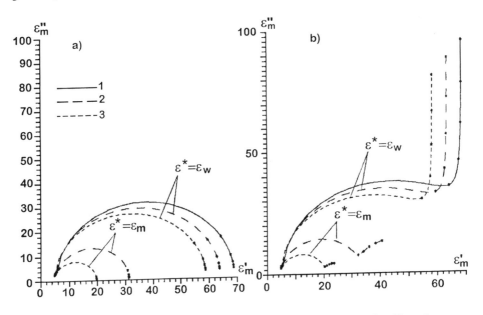

Figure 3.42. Complex effective permittivity of the air-bubbles-in-water emulsion. The volume content of air is: 1—$c = 0$; 2—$c = 0.05$, 3—$c = 0.1$. (a) Fresh water; (b) salt water.

The following features of the air–water mixture diagrams can be identified:

- Dielectric properties of two-phase media are changed with an increase in the bubbles' concentration in the mixture. Diagrams are shifted to a lower value of the effective permittivity.
- In the case of fresh water the diagrams will hold their shape, but in the case of salt water the shape is disrupted. In the $\lambda = 10$–30 cm wavelength range the right part of the diagram is streamed up.
- In the $\lambda = 0.1$–1 cm wavelength range the influence of the bubbles' concentration, temperature, and salinity of the water on a value of the effective permittivity is low.

The change in dielectric properties of the water skin-layer produces variations in microwave emission. In the simplest case of a smooth surface, the brightness temperature of a uniform medium with complex effective parameters ε_m is equal to $T_B = \left(1 - |\mathbf{r}(\varepsilon_m)|^2\right)T_0$, where $\mathbf{r}$ is the complex Fresnel reflection coefficient, and T_0 is temperature of the medium.

Modeling of the microwave emission spectra is shown in Figure 3.43. The brightness temperature is calculated for both limiting case, when the complex parameter is $\varepsilon^* = \varepsilon_0$; $\left\{A_1 = A_2 = A_3 = \frac{1}{3}\right\}$, and $\varepsilon^* = \varepsilon_m$; $\left\{A_1 = A_2 = 0, A_3 = 1\right\}$.

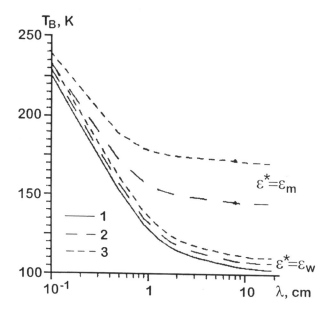

Figure 3.43. Spectra of the brightness temperature of the air-bubbles-in-water emulsion (at nadir).
1—$c = 0$; 2—$c = 0.05$; 3—$c = 0.1$.

In the first case (bubbles are spheres) the dependence of the brightness temperature on the air concentration is low, but in the second case (bubbles are disks) the dependence is strong. It is shown also that the microwave effects from the bubbles appear in a wide wavelength range of $\lambda = 1$–100 cm. The greater the wavelength of emission, the greater is the value of the brightness temperature contrast. So, the microwave signatures reflect primarily a variance of the bubbles' concentration in the subsurface ocean layer. In the $\lambda = 8$–21 cm wavelength range, variations in the brightness temperature due to aeration of water can reach about $\Delta T_B = 1$–15 K.

3.6 COMBINED FOAM–SPRAY–BUBBLES MODELS

The problem of the influence of dispersed media on the radiophysical properties of the ocean–atmosphere interface is a central problem in the study of ocean dynamics and surface disturbances under storm conditions using passive or active microwave remote sensing techniques. Therefore combined microwave models of two-phase ocean boundary layer should also be considered. The variant of the microwave model takes a hierarchy of dispersed oceanic structures into account (Figure 3.44).

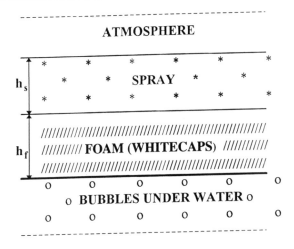

Figure 3.44. Combined model of two-phase ocean–atmosphere interface.

We choose the representation in the form of the layered non-uniform medium with vertically distributed parameters. In the common case a three-layer system can be considered: the upper first layer, bordering with the atmosphere, is the spray layer; the second layer is a foam (or whitecaps); the third layer is a population of bubbles, distributed in the water subsurface, below which the uniform water medium is located. To describe the action of the complex systems, different electrodynamic methods are applied. So, spray and aerosol are modeled by a discrete system of spherical water droplets; foam/whitecaps are modeled by a non-uniform layer with distributed effective dielectric parameters; and the subsurface bubbles population can be represented by the dielectric mixture of air and water.

The set of varied parameters is as follows:

- Temperature and salinity of the water.
- Function of the size distribution of spray (droplets).
- Water content of the spray or volume concentration of the water.
- Height of the spray layer.
- Function of the size distribution of the foam/whitecaps bubbles.
- Average thickness of the bubbles' shell.
- Concentration of bubbles in the foam/whitecaps layer.
- Thickness of the foam/whitecaps layer.
- Volume concentration of the underwater gaseous bubbles or subsurface bubbles' population.

The numerical algorithm is based on a combination of the scalar theory of radiation transfer in a discrete scattering system of water droplets like spray, the macroscopic theory of densely packed dispersed media like foam/whitecaps, and the models of matrix dielectric mixtures for analysis of effects from underwater bubbles' population. The modeling was made to conduct detailed investigations of the microwave emission of different types of dispersed structure in the $\lambda = 0.1$–10 cm wavelength range. As a result the following important features of microwave emission were manifested.

Microwave effects due to spray, which is located over the foam coverage, are different, because the spectral characteristics of the emission are essentially different for the cases of the foam coverage and free water surface. We calculated the brightness temperature contrast due to spray relative to foam coverage.

Figure 3.45 shows the spectral dependencies of the brightness contrast, which is determined by the polydispersed layer of spray *only*. The average thickness of the foam layer varies; the other parameters of the model are held fixed. In the case of the foam-free system, the brightness temperature contrast of the spray–water system ΔT_B is always positive.

However, when a foam layer is introduced in the model, the contrast can be both positive and negative. The spectrum of the brightness contrast $\Delta T_B(\lambda)$ is significantly

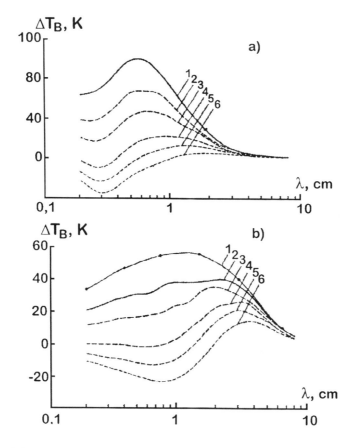

Figure 3.45. The brightness temperature contrast of the water–foam–spray system due to spray *only*. The thickness h of the foam layer is varied: 1—spray over water without foam ($h = 0$); 2—$h = 0.5$ cm; 3—$h = 0.1$ cm; 4—$h = 0.2$ cm; 5—$h = 0.3$ cm; 6—$h = 0.5$ cm. (a) Small-sized ($r_{max} \approx 0.01$ cm) spray; (b) large-sized ($r_{max} \approx 0.1$ cm) spray.

deformed owing to the change in the size distribution function of spray. For example, in the case of small-sized spray (Figure 3.45(a)) the main effects are observed in the short-wavelength range of $\lambda < 2$ cm; but in the case of large-sized spray the effects are observed in the wide-wavelength range of $\lambda = 0.2$–8 cm (Figure 3.45(b)). The value of the brightness contrast ΔT_B also depends on the other parameters of the model and, first of all, on the surface mass concentration of spray. The range of possible variations in the brightness contrast is $\Delta T_B = -40$ to $+100$ K in the case of small-sized spray, and $\Delta T_B = -20$ to $+60$ K in the case of large-sized spray.

The following types of two-phase microwave model can be suggested for ocean remote sensing applications:

Spray over the smooth water (spray + water)

- Microwave emission from the spray located above the smooth water surface is affected not only by scattering and absorption factors of individual spherical droplets, but also by the mass concentration of the spray. The emissivity of the spray becomes higher with increasing mass concentration. The spectrum of microwave emission is more sensitive to variations in size distribution and the concentration of spray.
- In the case of monodispersed spray (the size of droplets should be same) resonant effects in the emission are pronounced, while for polydispersed spray (the size of droplets is different) the resonant effects are smoothed out.
- Variations in the brightness temperature of the spray–water system can reach 30–60 K and higher in the $\lambda = 0.3$–2 cm wavelength range. This range will be optimal for indication of spray, or large-sized aerosol over the open (foam/whitecaps free) ocean surface.

Spray over the foam coverage (spray + foam + water)

- Introducing the model that includes the foam/whitecaps layer between the spray and water surface qualitatively changes the spectrum of microwave emission. The foam layer is a collection of spherical bubbles with certain microwave properties. So, the spectrum of the microwave emission is determined by the emissive features of the foam/whitecaps like a 'black body'. However, the effects connected with spray can yield both positive and negative brightness temperature contrasts. The observed 'surface cooling' effect due to the influence of the spray can be explained by specific scattering properties of water droplets in the microwave range.
- The combined influence of the foam–spray on the microwave emission is best shown on the $\lambda = 0.5$–2 cm wavelength range. In this range the effect like the resonant maximum is observed. The brightness temperature contrast in this range varies approximately from -20 K to $+40$ K.

The influence of bubble population (spray + foam + subsurface bubbles + water)

- A layer of gaseous bubbles, which is located between the foam (or foam + spray) layer and the smooth water surface, produces some changes in the spectrum of the microwave emission. At the same time, variations in the brightness temperature decrease. This is because the properties of the uniform water skin-layer are changed owing to the formation of a two-phase dielectric mixture. Some 'calming' of the spectrum of the microwave emission occurs.

• The effects of 'calming' and variations in the brightness temperature depend on the volume concentration of gaseous bubbles in the water. Increases in the bubble concentration decrease variations in the microwave emission.

In the case of a densely dispersed medium when the volume concentration of the water component is large ($c > 20\%$), the ocean–atmosphere interface at the range of wavelengths $\lambda = 0.8$–18 cm can be represented as a macroscopic electromagnetic system with distributed parameters. For example, a flat layering dielectric system is a reliable model for the estimation of spectral dependencies of the ocean microwave emission in stormy conditions. The model operates with vertical distribution of the effective permittivity that depends on the geometry of phase components and particles of the medium. The total reflection coefficient $r'_{n-1,n-2}$ and emission coefficient $\kappa_{n-1,n-2}$ of an n-layered system is calculated by using the following recursive procedure:

$$r'_{n-1,n-2} = \frac{r_{n-1,n-2} + r_{n,n-1}\exp(2i\psi_{n-1})}{1 + r_{n-1,n-2}r_{n,n-1}\exp(2i\psi_{n-1})}, \tag{3.39}$$

$$r'_{n-2,n-3} = \frac{r_{n-2,n-3} + r'_{n-1,n-2}\exp(2i\psi_{n-2})}{1 + r_{n-2,n-3}r'_{n-1,n-2}\exp(2i\psi_{n-2})},$$

$$\cdots\cdots\cdots\cdots\cdots\cdots\cdots\cdots\cdots\cdots\cdots$$

$$r'_{21} = \frac{r_{21} + r'_{32}\exp(2i\psi_2)}{1 + r_{21}r'_{32}\exp(2i\psi_2)},$$

$$\psi_n = 2\pi\frac{h_n}{\lambda}\sqrt{\varepsilon_n}\cos\theta_n, \qquad \sqrt{\varepsilon_{n-1}}\sin\theta_{n-1} = \sqrt{\varepsilon_n}\sin\theta_n,$$

$$\kappa_{n-1,n-2}(\lambda) = 1 - \left|r'_{n-1,n-2}(\lambda)\right|^2,$$

where $n \geqslant 3$ is the number of elementary flat layers; $r_{n-1,n-2}$ is the reflection coefficient (for any linear polarization) from the medium with number $n-2$ into the medium with number $n-1$; h_n is the thickness of the elementary layer with number n; θ_n is the angle of reflection from the layer with number n; ε_n is the complex permittivity of the elementary layer with number n.

Usually $n = 50$–100 is enough for the numerical modeling of dispersed media with non-trivial distribution of complex effective parameters. Variance of the law of vertical distribution of effective permittivity $\varepsilon_f(z)$, where z is the vertical coordinate, gives rise to variance of the emissivity spectrum at microwave frequencies. In practice, effects of interference between reflections from elementary layers are smoothed significantly due to spatial–temporal averaging of the microwave signal, associated with the antenna footprint and the effective band of radiometric receiver. Therefore it is possible to identify the type of non-uniformity of the dispersed subsurface or interface medium by using synchronized multi-frequency radiometric measurements when the spectrum of emissivity is defined well at the wide range of wavelengths. Usually, the multi-layer microwave model shows good possibilities at the retrieval of physical parameters of environment media when highly contrasting variations of the emissivity are observed. For example, it was shown [57, 58] that sea foam coverage and young sea ice are best suited to the multi-frequency diagnostics at wide range of the wavelengths $\lambda = 0.2$–21 cm.

3.7 A COMPLEX OCEAN MICROWAVE MODEL AND RADIATION–WIND DEPENDENCIES

The retrieval of wind-wave characteristics by using microwave radiometric measurements is the main question at the present time. Numerical algorithms and instruments [59–64] that have been developed for global observation of the oceans have serious limitations in their ability to retrieve the wind speed vector and the sea surface temperature under conditions of higher wind speeds in a fully developed storm ($V > 20$–25 m/s). Unlike empirical linear models and algorithms, the suggested complex microwave model of the ocean is designed on the basis of detailed physical modeling of microwave effects from each hydrophysical factor including multi-scale surface waves, dispersed structures and even specific surface disturbances induced by falling rain droplets [65]. The complex model gives strong and perhaps indefinite spectral dependencies of the ocean brightness temperature, associated with high microwave emission and dynamics of foam/whitecaps. These effects may be causing the non-linearity of the wind-vector retrieval algorithms, which have been used in particular for analysis of SSM/I data.

For the simplest case the contributions of factors on the total microwave emission of the ocean can be considered as statistically independent. Therefore, the total brightness temperature that is measured by a radiometer at a fixed wavelength and polarization is estimated as the sum [66]:

$$\Delta T_{\mathrm{B}}(\lambda, V) = \sum_{i=1}^{N} \Delta T_{\mathrm{B}i}(\lambda, V) W_i(V), \tag{3.40}$$

where $\Delta T_{\mathrm{B}i}$ is the spectral value of the brightness temperature of the factors; W_i are the corresponding weight coefficients that are equal to the area fractions (%) occupied by the factors i; V is the wind speed; and N is the number of factors. Note that the row (3.40) is an *a priori* non-linear function relative to the wind speed parameter V. The statistical information about the W_i-distribution can be found from additional empirical data, for example, from optical/radar remote sensing data. Also a large number of empirical functions $W(V)$ for both foam coverage and whitecaps in the ocean have been collected. The foam/whitecaps fraction W depends on the temperature and viscosity of the salt water, atmospheric stability of the ocean boundary layer, and the wind fetch also. The power law $W(V) = \alpha V^{\beta}$, where α and β are empirical constants, is used for modeling. However, this law is reliable only at wind speeds $V = 7$–20 m/s when the statistics and fraction of foam/whitecaps coverage are changed continuously. Such conditions are usually realized at the wind fetch. At fully developed storm conditions, as shown on aircraft optical measurements (see Chapter 2), function $W(V)$ reaches saturation and the value of the fraction W is constant. Therefore, our knowledge of the W-distribution law at different ocean surface conditions is an important element of the complex microwave model and the wind-vector retrieval algorithms. Spatial–temporal deviations of the $W(V)$-distribution, in particular the scattering of empirical coefficients α or/and β, are possible reasons for the large variability of the ocean microwave emission at wind speed $V > 7$–10 m/s. To reduce the systematic errors of wind speed retrieval, we need to apply non-linear algorithms and/or models. Recently, such an algorithm was designed on the basis of a neural network and was successfully applied to the analysis of the SSM/I satellite data. The algorithm essentially improves the precision of wind speed retrieval at stormy conditions [67, 68].

As an example, consider the results of the complex modeling of the brightness temperature contrast of the ocean in the wavelength range $\lambda = 0.2$–8.2 cm (at nadir). We use the model with the following hydrophysical factors: (1) large-scale linear gravity waves; (2) small-scale gravity–capillary surface waves; (3) foam streaks + spray; and (4) white-caps + spray. In this case, the sum (3.40) may be rewritten

$$T_B = (\kappa_{s1}W_s + \kappa_{f1}W_{f1} + \kappa_{f2}W_2)T_0, \tag{3.41}$$

where the emissivity due to the geometrical factor is

$$\kappa_{s1} = \iint \kappa_s(\theta; z_x, z_y) P_\theta(\theta; z_x, z_y)\, dz_x\, dz_y, \tag{3.41a}$$

$$\kappa_s = \kappa_{s0} + 2k_0^2 \iint G(K/k_0; \theta, \varphi) F(K, \varphi) K\, d\varphi\, dK,$$

and the emissivity due to the volume factor is

$$\kappa_{f1} = 1 - |r_{f1}|^2, \quad \kappa_{f2} = 1 - |r_{f2}|^2, \tag{3.41b}$$

and the statistical factors are

$$W_s + W_{f1} + W_{f2} = 1; \tag{3.41c}$$

$$W_{f1} = \frac{1}{1 + R_f W}; \quad W_{f2} = \frac{R_f}{1 + R_f W}; $$

$$R_f = A + BV; \quad W = \alpha V^\beta. \tag{3.41d}$$

Here: κ_{s1} is the contribution of the tilted small-scale resonant component of the surface waves and $P(\ldots)$ is the probability distribution of gravity wave slopes (according to the modified two-scale model); κ_{f1} is the combined contribution of foam streaks + spray; κ_{f2} is the combined contribution of whitecaps + spray. The effective reflection coefficients of the foam/whitecaps and spray κ_{f1} and κ_{f2} are defined from the electromagnetic models of dispersed media; R_f is the ratio of foam and whitecaps area fractions; A, B, α, β are empirical constants.

The total brightness temperature contrast is estimated as $\Delta T_B = T_B - T_{B0}$, where T_{B0} is the brightness temperature of smooth foam/whitecaps free water surface which depends on the wavelength and view angle (polarization). The models (3.40) and (3.41) describe radiation–wind dependencies from the ocean at the fixed wavelength and geometry of observation. The behavior of radiation–wind characteristics as well as values of the total brightness temperature contrast due to the combined influence of wind waves and foam/whitecaps/spray structures are determined by the W-distribution. At increasing wind speed when fully developed storm conditions are reached, function $W(V)$ should tend to a constant level, which will depend on the wind fetch value. Thus, the function $\Delta T(\lambda, V)$ will also tend to a constant level.

Figure 3.46 shows the three-dimensional view of the brightness temperature contrast (with respect to a smooth water surface) $\Delta T_B(\lambda, V)$. The increasing contrast with increasing wind speed is conditioned by the foam/whitecaps and spray influence, and the emissivity has a clear spectral dependence. Using the complex microwave model in general form (3.40) the various dynamical conditions of the ocean surface can be analyzed.

To test the complex model we carry out a comparison of some experimental and theoretical data. Figure 3.47 shows the calculated spectra of the radiation–wind sensitivity $\Delta T_B/\Delta V$ for different states of the ocean–atmosphere system. Calculations were made at several wavelengths: $\lambda = 0.8$, 1.35, 2.0, 3.2, 4.6, and 6.5 cm using the complex microwave model of the ocean [69]. Experimental data were collected in a wind speed range of $V = 0$–25 m/s [23, 70]. Atmospheric parameters like water vapor of the clouds and humidity are varied. Approximately two fields on the spectral dependence of the radiation–wind sensitivity are separated: at the wavelengths $\lambda = 0.8$–3.5 cm the influence of atmospheric parameters grows, and in the range of $\lambda \geqslant 4$ cm only ocean surface features are affected. It is important that the maximum of the radiation–wind sensitivity is located in the wavelength range about $\lambda = 3$–5 cm, where the influence of the atmosphere is not appreciable.

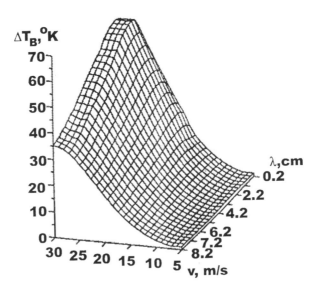

Figure 3.46. Example of the complex modeling of the ocean brightness temperature contrast $\Delta T_B(\lambda, V)$ vs. emission wavelength λ and wind speed V at nadir view angle.

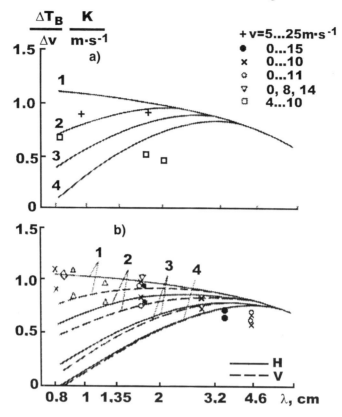

Figure 3.47. Spectra of the radiation–wind sensitivity in the ocean–atmosphere system. The view angle: $\theta = 0°$ (a) and $\theta = 50°$ (b). H, V are horizontal and vertical polarization. Atmosphere parameters: water vapour of the clouds are varied: 1—0; 2—0.71; 3—2.0; 4—3.5 kg/m^3; humidity is 14.9 kg/m^3 (constant). Symbols are the experimental data.

3.8 MICROWAVE DIAGNOSTICS OF OIL SPILLS

Surface films are frequently encountered in the ocean. Along with their origin, they can be divided into two categories: natural and artificial. Natural slicks are formed owing to the chemical and biological processes in the ocean surface; they are known by the name 'surface-active films'. Artificial slicks have appeared on the ocean surface as a result of the anthropogenic activity of human beings. These are polluting surface-active films of oil (or petroleum) and other synthetic and detergent oil products. They generally agglomerate in coastal economically advanced zones. The gross amounts of oil products in the ocean are estimated to be 6 to 12 million tons.

Surface-active films are manifest in the form of 'smooth surface' areas. The thickness of such films is several monomolecular layers ($\sim 10^{-7}$–10^{-6} cm). Organic surface-active films change the optical characteristics of the surrounding water. Sometimes their presence causes the appearance of anomalous phenomena in the reflected light due to the variance in the slopes of short gravity waves. The configuration of the slicks varies: they can be long streaks, oriented along the direction of the wind, or separate areas reminiscent of Langmuir cells. Surface-active films are optical indicators of oceanic internal waves.

For remote sensing applications, the most interesting presents the effects of the interaction of surface films and wind waves. The range of frequencies in which the variation of spectral density of the wave energy is observed can be quite wide: from 3 to 15 Hz [71]. Radar measurements also show that surface-active films sharply change the regimes of generation of gravity–capillary waves in the wind field.

Unlike surface-active films, oil films never make monomolecular layers. The range of thicknesses of typical oil films is $h = 10^{-4}$–1 cm. Thin films of crude oil give a silver trace; thicker oil films have a dark color without interference painting. Layers of water/oil emulsion remind the observer of a thick 'chocolate mousse'. The thickness of such layers can equal several centimeters.

Laboratory and natural experiments show the oil films effect on the process of wind-wave generation. On a wave crest the thickness of the film is usually more than that of a wave hollow. Polluting films depress the high-frequency components of surface wave spectrum much more than organic films Along with that, they slow down the mass-heat exchange between the ocean and atmosphere. In the slick area, the temperature of the ocean surface can increase by up to 1–2°C owing to the effects of the solar radiation absorption and screening.

In recent years the active and passive remote sensing methods of the detection of oil spills in the ocean surface have been developed. Two mechanisms of the influence of oil film on microwave emission can be considered: the attenuation of the high-frequency component of the wavenumber spectrum of surface waves, and the 'matching' action of a film on the microwave emission of the water surface.

Airborne microwave/radar remote sensing investigations of natural oil pollutions were made on the Caspian Sea [23]. It was found that polluted areas of the sea are observable as 'radio-shadows'. The effect is manifest in the form of negative radio-brightness contrasts from the oil films in the presence of non-stationary wind-wave conditions. The speed of the oil slick evolution in 'radio-shadows' areas approximately corresponds to the wind speed, while the size of these areas can reach several dozens of kilometers. Also it was found that spatial dynamics and time of the existence of oil slicks on the sea surface depend on hydrological features of the test region. These effects can be connected with a 'damping' factor.

The effect of impedance matching due to the oil film directly relates to appreciable differences in the complex dielectric constants of oil products (gasoline, benzene, petroleum) and sea water. Some experimental data on the dielectric properties of the oils and their derivatives are shown in Table 3.2. The value of the real part of the complex dielectric constant is equal to $\varepsilon' = 1.8$–3.0. Dielectric losses of the oil products are small. The tangent of dielectric losses is of the order of $\tan(\varepsilon'/\varepsilon'') = 10^{-3}$. Laboratory measurements show that ε' linearly depends on the specific weight of a purified oil and the time of its stay in open air. With the increase in temperature, the value ε' insignificantly subsides.

Table 3.2. Dielectric parameters of oil products

Oil type	ε'	$\tan \varepsilon'/\varepsilon''$	$t, °C$	Radiofrequency
Benzene	2.25–2.27	—	—	0.1–1 GHz; 5–10 GHz
Industrial benzene	2.10	$3 \cdot 10^{-3}$	—	35 GHz
Crude oil	2.12–2.25	10^{-3}	20–25	10 MHz; 3.9–10 GHz
Refined oil	1.8–3.0	$5 \cdot 10^{-3}$	23	37 GHz

Dielectric properties of water-in-oil emulsion differ essentially from the dielectric properties of the pure or raw oil. A simple microwave model of the film–water system is shown in Figure 3.48. The brightness temperature should vary with the change of parameters of emulsion and film thickness.

For calculation of the complex permittivity of the water-in-oil emulsion, formula (3.37) can be used. In this case we need to change the parameters: $\varepsilon_w \rightarrow \varepsilon_i$ and $\varepsilon_i \rightarrow \varepsilon_w$, where ε_i and ε_w are complex dielectric constants of oil and water; and A_j is the form-factor of water inclusions. Calculations show that typical values of the effective permittivity of emulsion are $\varepsilon'_m = 1.5$–5.0 and $\varepsilon''_m = 1.0$–3.5 in the $\lambda = 0.2$–2.0 wavelength range at the concentration of the water in emulsion $c = 0$–0.2. At large concentrations, the c influence of the form-factor A_j on the value of ε_m is essential.

The emissivity of thin film on the water surface can be calculated through the reflection coefficient of flat two-layer dielectric media [72]:

$$\kappa_m = 1 - \left| \frac{r_{12}\,e^{-2i\psi} + r_{23}}{r_{12}r_{23} + e^{-2i\psi}} \right|^2, \tag{3.42}$$

$$\psi = \frac{2\pi h}{\lambda}\sqrt{\varepsilon_m - \sin^2\theta},$$

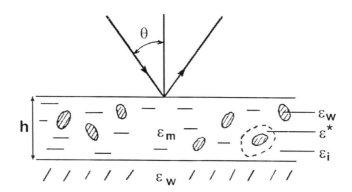

Figure 3.48. Microwave model of water-in-oil emulsion's film on the water surface.

where r_{12}, r_{23} are the coefficients of the reflection from the corresponding film boundaries; h is the thickness of a film; and θ is the angle of view.

The brightness contrast of the film–water system is $\Delta T_B = (\kappa_m - \kappa_0)T_0$; κ_0 is the emissivity of the oil-free water surface. In the $\lambda = 0.2$–0.8 cm wavelength range the contrast ΔT_B increases with the growth of the bulk concentration of water in the emulsion; it can reach $\Delta T_B = 60$–80 K. In the $\lambda > 2$ cm wavelength range the contrast is $\Delta T_B < 2$ K.

Figure 3.49 illustrates the typical interference dependencies of the brightness temperature of a two-layer dielectric system. The period of oscillations is estimated by

$$H = \frac{\lambda}{2\sqrt{\varepsilon'_m - \sin^2 \theta}} . \tag{3.43}$$

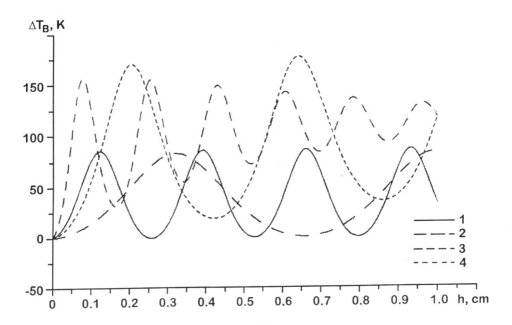

Figure 3.49. The dependence of the brightness contrast vs. thickness of oil film. Modeling at the wavelengths $\lambda = 0.8$ and 2 cm (at nadir). The bulk concentration of the water in emulsion: 1—$c = 0$ and $\lambda = 0.8$ cm; 2—$c = 0$ and $\lambda = 2$ cm; 3—$c = 0.5$ and $\lambda = 0.8$ cm; 4—$c = 0.5$ and $\lambda = 2$ cm.

In the case of a crude oil (without dielectric losses) the amplitude of oscillations is constant and is independent of the thickness of a film. But in the case of water-in-oil emulsion (the losses are introduced) the oscillations will attenuate, and the asymptotic level of the brightness temperature is determined by the complex dielectric constant of the film only.

Figure 3.50 shows the dependencies of the brightness temperature of the system on the view angle. The thickness of the film is varied. At the Bruster angle of view of about $\theta = 65–68°$, microwave effects do not appear on the vertical polarization, owing to film thickness, and the brightness temperature is defined by the dielectric constant of the oil medium only. So, by using the microwave measurements at these angles, it is possible to estimate the value of the dielectric constant of the oil. At the grazing angles of view $\theta = 70–80°$ on the vertical polarization, the increase in bulk concentration of water in emulsion causes a decrease in brightness temperature contrast.

Figure 3.51 illustrates microwave signatures of water-in-oil emulsion. Influence of the film thickness h and the bulk concentration of the water c on the brightness temperature contrast are both shown by a three-dimensional diagram $\Delta T_B(h, c)$. In the region of small values of thickness and concentration dependencies, $\Delta T_B(h)$ and $\Delta T_B(c)$ are well separated. The dependencies are linear in the region of small parameters of h and c.

Two-channel regression of the brightness contrasts due to influence of oil film, calculated on the wavelengths $\lambda = 0.8$ and 2 cm, is shown in Figure 3.52. The curves have a loop form and represent interference features of the reflection and emission from a two-layer dielectric medium. The reading of the oil thickness with the discrete of 50 micrometers is marked by the dots.

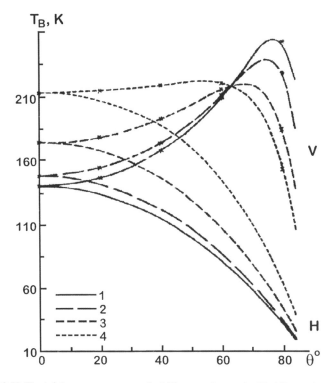

Figure 3.50. The brightness temperature of oil films vs. view angle. Modeling at the wavelength $\lambda = 0.8$ cm. Polarizations: vertical (V) and horizontal (H). The bulk concentration of the water in emulsion: 1—$c = 0$; 2—$c = 0.2$; 3—$c = 0.4$; 4—$c = 0.5$.

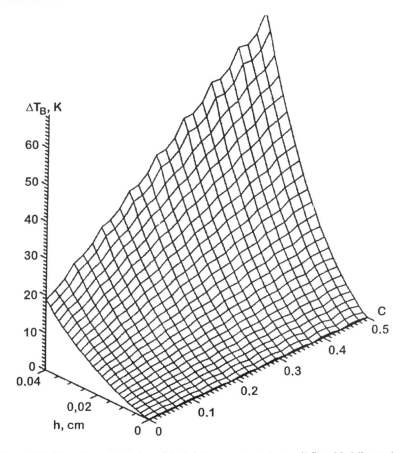

Figure 3.51. Three-dimensional view of the brightness contrast due to oil film. Modeling at the
wavelength $\lambda = 0.8$ cm (at nadir).

A comparison of the theoretical and experimental data is shown in Figure 3.53. We
use the results of airborne and laboratory microwave measurements of natural oil slicks.
Variations in the parameters of the model allow one to obtain the optimal agreement
between theory and experiments. In the case of airborne measurements at the wavelengths $\lambda = 0.8$ and 2 cm the best agreement occurs for the water-in-oil emulsion; in the
case of laboratory measurements at the wavelength $\lambda = 2$ cm it will be for the crude oil,
spreading on the smooth water surface.

By using multifrequency radiometry it is possible to measure the thickness of oil
pollution and phase composition of emulsion. The combination of multifrequency microwave and radar remote sensing techniques permits estimation of the volume of spreading
oil, and sometimes the determination of the concentration of the polluting substances.
The $\lambda = 0.3$–2.0 cm wavelength range is the optimal range for passive microwave diagnostics of oil pollutions in the ocean. Passive microwave measurements should be made
using both polarizations at the view angle of $\theta = 0$–$30°$.

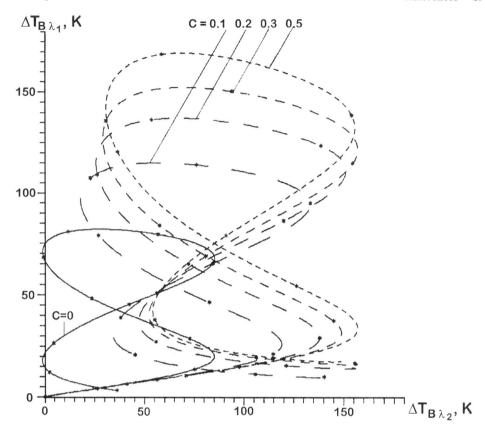

Figure 3.52. Two-channel regression of the brightness temperature contrasts due to oil film at the wavelengths $\lambda_1 = 2$ cm and $\lambda_2 = 0.8$ cm (at nadir). The bulk concentration c of the water in emulsion is changed from 0 to 0.5.

REFERENCES

[1] Ulaby, F.T., Moore, R.K., and Fung, A.K. (1981, 1982, 1986) *Microwave Remote Sensing. Active and Passive* (in three volumes). N.Y.: Artech House.

[2] Tsang, L., Kong, J.A., and Shin, R.T. (1985) *Theory of Microwave Remote Sensing.* N.Y.: Wiley-Interscience.

[3] Fung, A.K. (1994) *Microwave Scattering and Emission Models and Their Applications.* Boston: Artech House, 573 pp.

[4] *Atmospheric Remote Sensing by Microwave Radiometry* (1993). M.A. Janssen (ed.). N.Y.: Wiley.

[5] Hasted, J.B. (1961) 'The dielectric properties of water', *Progress in Dielectrics*, **3**, 103–149.

[6] Stogryn, A. (1971) 'Equations for calculating the dielectric constant for saline water', *IEEE Trans. Microwave Theory and Technology*, **19**, No. 8, 733–736.

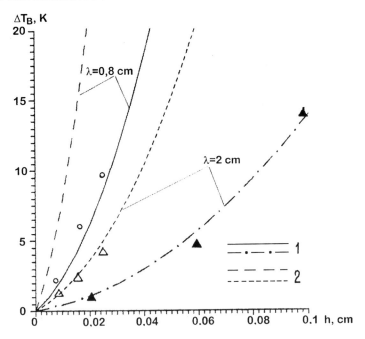

Figure 3.53. The dependence of the brightness temperature contrast on thickness of oil films at the wavelengths $\lambda = 0.8$ and 2 cm (at nadir). Calculations: 1—crude oil; 2—water-in-oil emulsion ($c = 0.5$). Experimental data: o, Δ—airborne; ▲—laboratory.

[7]   Ray, P.S. (1972) 'Broadband complex refractive indices of ice and water', *Appl. Opt.*, **11**, No. 8, 1836–1844.

[8]   Ho, W. and Hall, W.F. (1973) 'Measurements of the dielectric properties of sea water and NaCl solutions at 2.65 GHz', *J. Geophys. Res.*, **78**, No. 27, 6301–6315.

[9]   Rayzer, V.Yu., Sharkov, V.Ye., and Etkin, V.S. (1975) 'Influence of temperature and salinity on the radio emission of a smooth ocean surface in the decimeter and meter bands', *Izvestiya, Atmospheric and Oceanic Physics*, **11**, No. 6, 652–655 (translated from Russian).

[10] Sharkov, E.A. (1984) 'To the question about relaxation model of dielectric properties of concentrated electrolytes', *Journal of Physical Chemistry*, **58**, No. 7, 1705–1710 (in Russian).

[11] Swift, C.T. and MacIntosh, R.E. (1983) 'Considerations for microwave remote sensing of ocean–surface salinity', *IEEE Trans. Geosci. Remote Sensing*, **21**, 480–490.

[12] Shutko, A.M. (1986) *Microwave Radiometry of A Water Surface and The Ground.* Moscow: Nauka, 189 pp. (in Russian).

[13] Hollinger, J.P. (1971) 'Passive microwave measurements of sea surface roughness', *IEEE Trans. Geosci. Electron.*, **9**, No. 3, 165–169.

[14] Swift, C.T. (1974) 'Microwave radiometric measurements of the Cape Cod Canal', *Radio Science*, **9**, No. 7, 641–653.

[15] Wu, S.T. and Fung, A.K. (1972) 'A noncoherent model for microwave emissions and backscattering from the sea surface', *J. Geophys. Res.*, **77**, No. 30, 5917–5929.

[16] Cox, T.S. and Munk, W. (1954) 'Statistics of the sea surface derived from sun glitter', *J. Marine Res.*, **13**, 198–227.

[17] Bass, F.G. and Fucks, I.M. (1979) *Wave Scattering from Statistically Rough Surfaces* Elmsford, N.Y.: Pergamon, 525 pp.

[18] Ishimary, A. (1991) *Electromagnetic Wave Propagation, Radiation, and Scattering.* Englewood Cliffs, New Jersey: Prentice Hall, 637 pp.

[19] Rytov, S.M., Kravtsov, Yu.A., and Tatarskii, V.I. (1989) *Principles of Statistical Radiophysics, Vol. 3*, Berlin: Springer-Verlag.

[20] Kravtsov, Yu.A., Mirovskaya, Ye.A., Popov, A.Ye., Troitskiy, I.A., and Etkin, V.S. (1978) 'Critical effects in the thermal radiation of a periodically uneven water surface', *Izvestiya, Atmospheric and Oceanic Physics*, **14**, No. 7, 522–526 (translated from Russian).

[21] Irisov, V.G., Trokhimovskii, Yu.G., and Etkin, V.S. (1987) 'Radiothermal spectroscopy of the ocean surface', *Sov. Phys. Dokl.*, **32**, No. 11, 914–915 (translated from Russian).

[22] Irisov, V.G. (1991) 'Electromagnetic model for rough surface microwave emission and reconstruct ripple spectrum parameters', *IGARSS'91 Conference. Proceedings.* Helsinki, Finland, pp. 1271–1273.

[23] Etkin, V.S., Raev, M.D., Bulatov, M.G., *et al.* (1991) *Radiohydrophysical Aerospace Research of Ocean (Part 1); Radio Hydrophysical Ecological and Hydrological Research. Instruments and Techniques for Data Processing (Part 2).* Academy of Sciences U.S.S.R., Space Research Institute, Moscow, Preprint No. 1749, 84 pp. (part 1) and 47 pp. (part 2).

[24] Pospelov, M.N. (1996) 'Surface wind speed retrieval using passive microwave polarimetry: The dependence on atmosphere stability', *IEEE Trans. Geosci. Remote Sensing*, **34**, No. 5, 1166–1171.

[25] Voronovich, A. (1994) 'Small-slope approximation for electromagnetic wave scattering at a rough interface of two dielectric half-spaces', *Waves in Random Media*, **4**, 337–367.

[26] Voronovich, A. (1996) 'On the theory of electromagnetic waves scattering from the sea surface at low grazing angles', *Radio Science*, **31**, No. 6, 1519–1530.

[27] Irisov, V.G. (1997) 'Small-expansion for thermal and reflected radiation from a rough surface', *Waves in Random Media*, **7**, 1–10.

[28] Il'in, V.A., Naumov, A.A., Rayzer, V.Yu., Filonovich, S.R., and Etkin, V.S. (1985) 'Influence of short gravity waves on the thermal radiation from the surface of water', *Izvestiya, Atmospheric and Oceanic Physics,* **21**, No. 1, 59–63 (translated from Russian).

[29] Il'in, V.A., Kamenetskaya, M.S., Rayzer, V.Yu., Fatykhov, K.Z., and Filonovich, S.R. (1988) 'Radiophysical studies of nonlinear surface waves', *Izvestiya, Atmospheric and Oceanic Physics*, **24**, No. 6, 467–471 (translated from Russian).

[30] Ilyin, V.A. and Raizer, V.Yu. (1992) 'Microwave observations of finite-amplitude water waves', *IEEE Trans. Geosci. Remote Sensing*, **30**, No. 1, 189–192.

[31] Su, M.-Y. (1982) 'Three-dimensional deep-water waves. Part 1. Experimental

measurement of skew and symmetric wave patterns', *J. Fluid Mech.*, **124**, No. 1, 77–108.

[32] Gershenzon, V.E., Raizer, V.Yu., and Etkin, V.S. (1982) 'The transition layer method in the problem of thermal radiation from rough surface', *Radiophysics and Quantum Electronics,* **25**, No. 11, 914–918 (translated from Russian).

[33] *Electromagnetic Theory of Gratings* (1980). R. Petit (ed.). Berlin, Heidelberg, New York: Springer Verlag, 284 pp.

[34] Kuz'min, A.V. and Raizer, V.Yu. (1991) 'Application of the theory of excursions of a random field to the analysis of radiation from a rough surface in the quasistatic approximation', *Radiophysics and Quantum Electronics*, **34**, No. 2, 128–135 (translated from Russian).

[35] Khusu, A.P., Vitenberg, Yu.R., and Pal'mov, V.A. (1975) *Roughness of Surface: Probability Theory Approach.* Moscow: Nauka (in Russian).

[36] Bolotnikova, G.A., Irisov, V.G., Raizer, V.Yu., Smirnov, A.I., and Etkin, V.S. (1994). 'Variations of the natural emission of the ocean in the 8 and 18 cm bands', *Soviet Journal of Remote Sensing*, **11**(3), 393–404 (translated from Russian).

[37] Droppleman, J.D. (1970) 'Apparent microwave emissivity of sea foam', *J. Geophys. Res.*, **75**, No. 3, 696–698.

[38] Rozenkranz, P.V. and Staelin, D.H. (1972) 'Microwave emissivity of ocean foam and its effect on nadiral radiometric measurements', *J. Geophys. Res.*, **77**, No. 33, 6528–6538.

[39] Stogryn, A. (1972) 'The emissivity of sea foam at microwave frequencies', *J. Geophys. Res.*, **77**, No. 9, pp. 1698–1666.

[40] Williams, G. (1971) 'Microwave emissivity measurements of bubble and foam', *IEEE Trans. Geosci. Electr.*, Vol. GE-9, No. 4, pp. 221–224.

[41] Smith, P.M. (1988) 'The emissivity of sea foam at 19 and 37 GHz', *IEEE Trans. Geosci. Remote Sensing*, **29**, No. 5, 541–547.

[42] Bordonskiy, G.S., Vasil'kova, I.B., Veselov, V.M., Vorsin, N.N., Militskiy, Yu.A., Mirovskiy, V.G., Nikitin, V.V., Rayzer, V.Yu., Khapin, Yu.B., Sharkov, Ye.A., and Etkin, V.S. (1978) 'Spectral characteristics of the emissivity of foam formations', *Izvestiya, Atmospheric and Oceanic Physics*, **14**, No. 6, 464–469 (translated from Russian).

[43] Militskii, Yu.A., Raizer, V.Yu., Sharkov, E.A., and Etkin, V.S. (1977) 'Scattering of microwave radiation by foamy structures', *Radio Engineering and Electronic Physics*, **22**, No. 11, 46–50 (translated from Russian).

[44] Rayzer, V.Yu. and Sharkov, E.A. (1981) 'Electrodynamic description of densely packed dispersed system', *Radiophysics and Quantum Electronics*, **24**, No. 7, 553–557 (translated from Russian).

[45] Van de Hulst, H. (1957) *Light-Scattering by Small Particles*. N.Y.: Wiley.

[46] Odelevskiy, V.N. (1951) 'Calculations of the general conductivity of heterogeneous layers', *Journal of Technical Physics*, **21**, No. 6, 667–685 (in Russian).

[47] Vorsin, N.N., Glotov, A.A., Mirovskiy, V.G., Raizer, V.Yu., Troitskii, I.A., Sharkov, E.A., and Etkin, V.S. (1984) 'Direct radiometric measurements of sea foam', *Sov. J. Remote Sensing*, **2**(3), 520–525 (translated from Russian).

[48] Bortkovskii, R.S. (1987) *Air–sea exchange of heat and moisture during storms.* Dordrecht, Holland: D. Reidel.

[49] Wu, J. (1990) 'On parametrization of sea spray', *J. Geophys. Res.*, **95**, No. C10, 18269–18279.

[50] Dombrovskiy, L.A. and Raizer, V.Yu. (1992) 'Microwave model of a two-phase medium at the ocean surface', *Izvestiya, Atmospheric and Oceanic Physics*, **28**, No. 8, 650–656 (translated from Russian).

[51] Dombrovskiy, L.A. (1972) 'Calculation of radiation heat transfer in a plane-parallel layer of an absorbing and scattering media', *Izvestiya Acad. Nauk SSSR, Mech. Zhid. I Gaza*, **4**, 165–169 (in Russian).

[52] Cherny, I.V. and Sharkov, E.A. (1988) 'Radio remote sensing investigations of sea wave breaking process', *Issledovaniya Zemli is Kosmosa*, No. 2, pp. 17–28 (in Russian).

[53] Medwin, H. (1977) 'In situ acoustic measurements of microbubbles at sea', *J. Geophys. Res.*, **82**, 971–976.

[54] Farmer, D.M. and Lemone, D.D. (1984) 'The influence of bubbles on ambient noise in the ocean at high speed', *J. Phys. Oceanogr.*, **14**, 1762–1778.

[55] Mulhearn, P.J. (1981) 'Distribution of microbubbles in coastal water', *J. Geophys. Res.*, **86**, N9. C7, 6429–6434.

[56] De Loor, G.P. (1983) 'The dielectric properties of wet materials', *IEEE Trans. Geosci. Remote Sensing*, **21**, No. 3, 364–369.

[57] Webster, W.J., Wilheit, T.T., Ross, D.B., and Gloersen, P.G. (1976) 'Spectral characteristics of the microwave emission from a wind-driven foam-covered sea', *J. Geophys. Res.*, **81**, 3095.

[58] Raizer, V.Yu., Zaitseva, I.G., Aniskovich, V.M., and Etkin, V.S. (1986) 'Determining sea ice physical parameters from remotely sensed microwave data in the 0.3–18 cm range', *Sov. J. Remote Sensing*, **5**(1), 29–42 (translated from Russian).

[59] Wilheit, T.T. (1979) 'A model for the microwave emissivity of the ocean's surface as a function of wind speed', *IEEE Trans. Geosci. Electron.*, **17**, No. 4, 244–249.

[60] Hollinger, J.P., Peirce, J.L., and Poe, G.A. (1990) 'SSM/I instrument evaluation', *IEEE Trans. Geosci. Remote Sensing*, **28**, No. 5, 781–790.

[61] Wentz, F.J. (1992) 'Measurement of oceanic wind vector using satellite microwave radiometers', *IEEE Trans. Geosci. Remote Sensing*, **30**, 25535–25551.

[62] Scou, N. (1989) *Microwave radiometer systems. Design & Analysis.* Artech House.

[63] Liu, W.T., Tang, W., and Wentz, F.J. (1992) 'Perceptible water and surface humanity over the global oceans from Special Sensor Microwave Imager and European Center for Medium Range Weather Forecasts', *J. Geophys. Res.*, **97**, 2251–2264.

[64] Goodberlet, M.A., Swift, C.T., and Wilkerson, J.C. (1989) 'Remote sensing of ocean surface wind with the Special Sensor Microwave/Imager', *J. Geophys. Res.*, **94**, No. C10, 14547–14555.

[65] Il'in, V.A., Kasymov, S.S., Rayzer, V.Yu., Stepanishceva, M.N., and Fatykhov, K.Z. (1991) 'Laboratory studies of disturbances on a surface caused by falling rain', *Izvestiya, Atmospheric and Oceanic Physics*, **27**, No. 5, 399–402 (translated from Russian).

[66] Raizer, V.Yu. (1992) 'Two phase ocean surface structures and microwave remote sensing', *IGARSS'92 Conference Proceedings*. Huston, TX, USA, Vol. 3, pp. 1460–1462.

[67] Krasnopolsky, V.M., Breaker, L.C., and Gemmill, W.H. (1995) 'A neural network as a nonlinear transfer function model for retrieving surface wind speeds from the special sensor microwave imager', *J. Geophys. Res.*, **100**, 11033–11045.

[68] Krasnopolsky, V.M., Gemmill, W.H., Breaker, L.C., and Raizer, V.Yu. (1997) 'An SSM/I wind speed retrieval algorithm with improved performance at higher wind speeds' (to be published).

[69] Kosolapov, V.S. and Raizer, V.Yu. (1991) 'Satellite microwave radiometry of the rain intensity and cloud water content (from modeling results)', *Sov. J. Remote Sensing*, **8**(5), 860–878 (translated from Russian).

[70] Sasaki, Y., *et al.* (1987) 'The dependence of sea-surface microwave emission on wind speed, frequency, incident angle and polarization over the frequency range from 1 to 40 GHz', *IEEE Trans. Geosci. Remote Sensing*, **25**, No. 2, 138–146.

[71] Huhnerfuss, H. and Walter, W. (1987) 'Attenuation of wind waves by monomolecular sea slicks', *J. Geophys. Res.*, **92**, No. C4, 3961–3963.

[72] Landau, L.D. and Lifshitz, E.M. (1984) *Electrodynamics of Continuous Media*, 2nd edn with L.P. Pitaevskii. Oxford, New York: Pergamon Press.

4

Instruments and methods for microwave remote sensing of ocean–atmosphere system

Microwave remote sensing of the ocean–atmosphere system is based mainly on using the frequencies of oxygen and water vapor absorption lines, as well as single frequencies located in the transparency windows of the atmosphere to determine the standard meteorological variables of the atmosphere and ocean surface. The aerospace techniques were developed for near nadir and moderate viewing angles [1–7].

In this section, we first describe the multichannel microwave radiometer for remote sensing of the ocean surface under grazing angles in the 20–100 GHz wave band and discuss the theoretic approach of observations under grazing angles. The application of multispectral microwave remote sensing of the ocean including non-typical frequencies lives up to expectations, because it allows us to observe many oceanic processes that are not available from other means (see Chapter 5). In addition, the increase in the viewing angle leads to a greater swath width of the aerospace technique, but in this case the refraction in the atmosphere must be taken into account.

Below, an airborne multichannel microwave imaging radiometer and the unique procedure of its in-flight absolute calibration are described. We then present the salient characteristics of the MTVZA satellite microwave radiometer, operating at frequencies from 19 to 183 GHz. This instrument is planned to be deployed on the 'Meteor-3M' spacecraft and will first be launched by the Russian Space Agency in August, 1999. A comparison analysis between satellite microwave radiometers of existing spacecraft DMSP (Defense Meteorological Satellite Program, USA) and 'Meteor-3M' is presented.

4.1 AIRBORNE MULTICHANNEL MICROWAVE IMAGING RADIOMETER

The airborne scanning multichannel radiometer provides the two-polarization measuring of radiation at frequencies of 22.2, 31, 34, 37, 42, 48, 75 and 96 GHz. A number of operating frequencies have been chosen experimentally. We developed the multichannel microwave system for the remote sensing of the ocean step by step, adding and testing the new operating frequencies in field experiments [8–10]. First, the microwave instrument was tested in shipboard experiments during 1980–1986. Since 1988, the multichannel microwave radiometer has been deployed on research aircraft in scanning mode. Figure 4.1 shows the block-diagrams of its latest version.

The main goal for creation of the multichannel microwave system for remote sensing of the sea surface was to coincide in time and space spectral and polarization measurements. This problem has been solved by using a multifrequency horn antenna. The broadband, ten-port antenna consists of two parts (see Figure 4.1). Part *1* is a conical horn with a sequence of waveguide tapers and orthomode transducers at the throat. It has ten waveguide output ports corresponding to the vertical and horizontal polarization signals at five frequency bands. Each waveguide output is located at the circular section of a conical horn, which is defined by the following condition for an electromagnetic wave of TE_{10} type [11]:

$$\lambda_c = 3.413r$$

where λ_c is the critical wavelength, and r the radius of the waveguide's critical section of conical horn.

Part *2* is a conical horn which forms the antenna pattern. It can be used as both the aperture antenna and the feed-horn of a parabolic-reflector antenna. The shape of the main beam of the antenna pattern is defined by the aperture d of the horn and angle of cone opening 2β.

According to [12], the aperture d of the feed-horn for the parabolic-reflector antenna is defined as

$$d = 2\frac{-\left(L+\frac{2}{n}\lambda\right)\mathrm{tg}\frac{\beta}{2}+\left[\left(L+\frac{2}{n}\lambda\right)^2\mathrm{tg}^2\frac{\beta}{2}+\frac{n}{2}\lambda\left(1-\mathrm{tg}^2\frac{\beta}{2}\right)\left(2L+\frac{n}{2}\lambda\right)\right]^{1/2}}{1-\mathrm{tg}^2\frac{\beta}{2}},$$

where L is the distance between the feed-horn and the parabolic-reflector; $n = 1; 1.5; 2;...$
$n = 1, 3,...$ corresponds to the cupola shape of the top of the antenna main beam, $n = 1.5, 2.5,...$ corresponds to the flat shape of the top of the antenna main beam, $n = 2, 4,...$ corresponds to the funnel shape of the top of the antenna main beam.

If the horn is used as an antenna, d is defined as

$$d = n\lambda\mathrm{ctg}\frac{\beta}{2}.$$

The design of the instrument is based upon Dicke-type radiometric principles. All channels are direct amplification radiometers, except the channels of 75 and 96 GHz. They are built as superheterodyne receivers using balanced mixers [9]. The direct amplification radiometers made of low-noise amplifier (LNA) are characterized by absolute stability in operating at multichannel systems. The amplification of LNA is 54–57 dB. LNA is a 7- to 9-stage amplifier, depending on operating frequency, and built on Field effect transistors with a 0.35-micron gate length. The bandwidth of LNA is 1–2 GHz. The receiver noise temperature for each frequency is shown in Figure 4.1. The square law detector is a Schottky-barrier diode. The detected signal passes through an LF amplifier

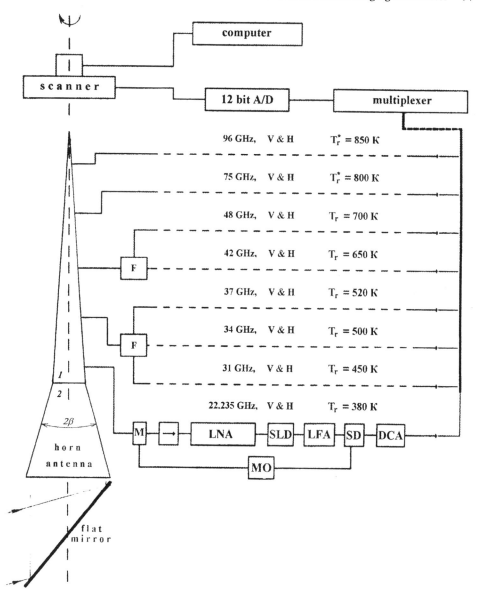

Figure 4.1. Block-diagram of airborne scanning multichannel microwave radiometer: F—filter; M—modulator; LNA—low-noise amplifier; SLD—square law detector; LFAA—low-frequency amplifier; SD—synchronous detector; DCA—direct current amplifier; MO—modulation oscillator.

and then converted by a synchronous detector to DC voltage. The output of the DC amplifier is fed into an array of multiplexers and a 12-bit analog-to-digital converter.

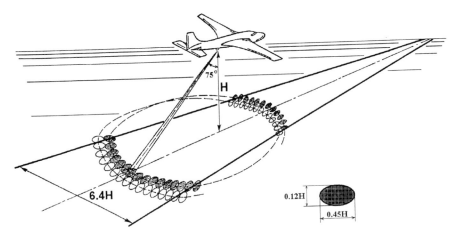

Figure 4.2. Airborne microwave imaging configuration.

Imaging configuration is based on a circular conical scanning with a viewing angle of 75° from nadir. To maintain the invariance of viewing angle and polarization in scanning sector the radiometer, horn-antenna and flat mirror are continually rotated about the vertical axis. The scanner contains the coaxially mounted bearing and signals and power transfer assembly. The airborne microwave imaging configuration is shown in Figure 4.2.

The data processing system built on a computer performs the complex functioning control, acquisition, storing, preliminary processing and displaying of data in real-time mode.

The parameters of the airborne multichannel microwave imaging technique, deployed on a TU-134 research aircraft, are

— angle of antenna beam incidence: 75°
— scanning sector (conical scanning in front and back hemisphere): 130°
— scanning period: 30 s
— half-power beamwidth (75 GHz): 2°
— antenna footprint (75 GHz): $0.12H \times 0.45H$ (where H is the flight height)
— swath width: $6.4H$

An unusual procedure for radiometer absolute inflight calibration is used. Because the remote sensing is run under grazing angles, the calibration procedure is performed while the aircraft is in a banked turn. In this case the radiation of two areas of known (calculated) brightness temperature are measured: the hot radio horizon and the cold sky above the horizon. The calibration procedure is discussed in section 4.3.

4.2 PECULIARITIES OF OBSERVATIONS UNDER GRAZING ANGLES (CALCULATION MODEL)

The aerospace remote sensing of the ocean–atmosphere system under grazing angles must take into account the spherical form of the Earth and refraction in the atmosphere.

Let us consider the microwave emission of the sphere-laminar atmosphere at height H from the underlying surface (Figure 4.3). There are three types of antenna beam trajectory. Calculation of the brightness temperature depends on the type of antenna beam trajectory [13].

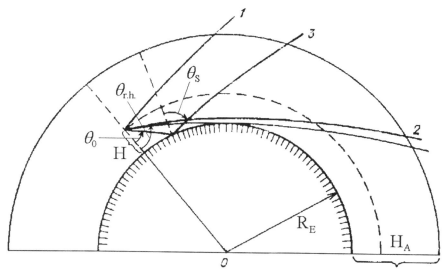

Figure 4.3. Three types of antenna beam trajectory for spherical-laminar atmosphere at normal refraction. θ_0—viewing angle, $\theta_{r.h.}$—angle of radiohorizon direction, θ_s—incident angle, R_E—radius of the Earth, H_A—height of atmosphere, H—viewing height.

When $90° \leqslant \theta_0$ (type 1, Figure 4.3), T_b is described by

$$T_{b\downarrow}(\theta_0, H) = \int_H^{H_A} \frac{\alpha(z)T(z)\, e^{-\tau(z,H_A)}\, dz}{\eta'(\theta_0, z)} + T_r\, e^{-\tau(H,H_A)}. \tag{4.1}$$

When $\theta_{r.h.} \leqslant \theta_0 < 90°$ (type 2, Figure 4.3), T_b is described by

$$T_{b\uparrow\downarrow}(\theta_0, H) = \int_{z_m}^{H} \frac{\alpha(z)T(z)\, e^{-\tau(z,H)}\, dz}{\eta'(\theta_0, z)} + \int_{z_m}^{H_A} \frac{\alpha(z)T(z)\, e^{-\tau(z,H_A)}\, dz}{\eta'(\theta_0, z)}\, e^{-\tau(z_m, H)}. \tag{4.2}$$

When $\theta_0 < \theta_{r.h.}$ (type 3, Figure 4.3), T_b is described by

$$T_{bp\uparrow}(\theta_0, H) = \kappa_p(\theta_s)T_s\, e^{-\tau(0,H)} + \int_0^{H} \frac{\alpha(z)T(z)\, e^{-\tau(z,H)}\, dz}{\eta'(\theta_0, z)}$$

$$+ \left(1 - \kappa_p(\theta_s)\right)\int_0^{H_A} \frac{\alpha(z)T(z)\, e^{-\tau(z,H_A)}\, dz}{\eta'(\theta_0, z)}\, e^{-\tau(0,H)}, \tag{4.3}$$

where

$$\theta_{\text{r.h.}} = \arcsin\left(\frac{R_E \cdot m'(0)}{(R_E + H) \cdot m'(H)}\right)$$

$\alpha(z)$ is the atmosphere absorption coefficient, according to [14]; $T(z)$ is the atmosphere temperature profile; T_s is the sea surface temperature; $T_r = 2.73$ K is the cosmic background temperature; κ_p is the sea surface emissivity, according to [15]; m' is the refraction of air index, according to [16]; $p = v,h$–polarization; z_m is the minimum distance between surface and antenna beam trajectory of type 2;

$$\tau(a,b) = \int_a^b \frac{\alpha(z)\,dz}{\eta'(\theta_0, z)} \text{ is the optical density of the atmosphere;}$$

$$\theta_s = \arcsin\left(\frac{m'(0) \cdot (R_E + H)}{m'(R_E) \cdot R_E} \sin\theta_0\right)$$

$$\eta'(\theta_0, z) = \sqrt{1 - \left(\frac{m'(0) \cdot (R_E + H)}{m'(z)(R_E + z)} \sin\theta_0\right)^2}.$$

Equations (4.1)–(4.3) were used for the calculation of the brightness temperature of scenes to calibrate and validate the experimental microwave data, obtained by means of shipboard and airborne radiometers (see Chapter 5).

4.3 ABSOLUTE IN-FLIGHT CALIBRATION OF AIRBORNE RADIOMETER

In microwave radiometry, hot and cold black-body targets of known brightness temperature are used to calibrate microwave radiometers. Below, we discuss the procedure of absolute in-flight calibration of airborne multichannel microwave radiometers described in section 4.1. This procedure is based on the measuring of two known scenes (calculated) brightness temperatures—the radiohorizon and the cold sky—when an aircraft banks over a water surface.

When the aircraft banks more than 25°, the antenna beam looks consecutively through the hot radiohorizon and the cold sky (Figure 4.4). The experimental distributions of brightness temperature T_b for frequencies for 22 GHz (horizontal polarization) and 75 GHz (vertical polarization) in the scanning sector when an aircraft banks 25° are shown in Figure 4.5. The maximum of T_b corresponds to viewing the radiohorizon, but the minimum of T_b corresponds to the cold sky. The experimental data are based on the ground calibration procedure of the radiometer, including hot black body and cold target with liquid nitrogen.

The brightness temperature of calibration scenes is calculated according to equations (4.1)–(4.3), based on the following: U.S. standard atmosphere model; spherical form of the Earth and refraction in the atmosphere. In addition, a 'polarization coupling' effect, an antenna pattern smoothing, scanning configuration parameters and aircraft navigation data are taken into account.

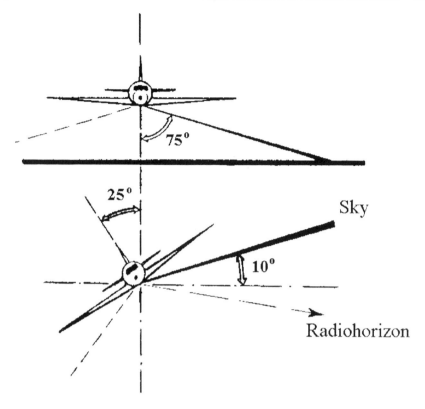

Figure 4.4.

'Polarization coupling' effect

The polarization basis of the instrument is rotated with respect to the surface basis in the scanning sector during banking of the aircraft. In this case, each radiometer channel receives the sea surface radiation of both vertical and horizontal polarization [17]:

$$\overline{T_b^i} = \begin{pmatrix} T_{bv^i} \\ T_{bh^i} \end{pmatrix} = \begin{pmatrix} \cos^2 \Psi & \sin^2 \Psi \\ \sin^2 \Psi & \cos^2 \Psi \end{pmatrix} \begin{pmatrix} T_{bv} \\ T_{bh} \end{pmatrix} = \overline{\overline{U}}(\Psi) \cdot \overline{T_b}, \tag{4.4}$$

where T_b is the vector of real polarization basis of sea surface; T_b^i is the vector of instrument polarization basis; Ψ is the angle of polarization basis rotation.

When an antenna beam looks through the radiohorizon, Ψ is defined by

$$\text{tg}\Psi = \cos\varphi \left(1 + \frac{\sin\varphi + \sin\alpha \cdot A}{\cos\alpha \cdot \text{tg}\theta_0 \sqrt{\cos^2\varphi + A^2} - \sin\alpha \cdot A} \right)$$

$$A = -\frac{\sin\alpha}{\text{tg}\theta_{r.h.}} + \sin\varphi \cdot \cos\alpha;$$

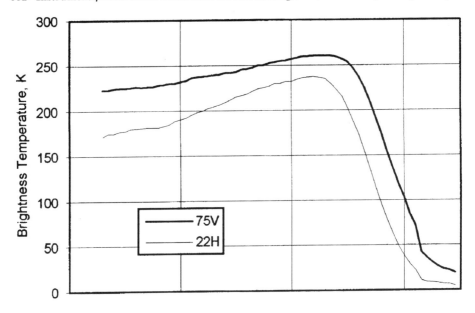

Figure 4.5. Experimental brightness temperature distribution in scanning sector according to Figure 4.4. Flight height is 10 km over the ocean surface. Clear atmosphere, sea surface temperature 14°C, absolute humidity 9.5 g/m³. Brightness temperature maximum corresponds to viewing to radiohorizon, but minimum cold sky.

$$\sin \varphi = \frac{1}{\mathrm{tg}\,\theta_{\mathrm{r.h.}} \cdot \mathrm{tg}\,\alpha} \left(\cos 2\theta_0 + \sin 2\theta_0 \sqrt{\frac{1}{4\cos^2 \alpha \cdot \cos^2 \theta_0 \cdot \sin^2 \theta_0} - 1} \right)$$

where α is the angle of bank of the aircraft, and $\theta_0 = 75°$ the angle of scanning cone opening.

Antenna pattern smoothing

The brightness temperature received by an antenna with aperture d is described by

$$\tilde{T}_{\mathrm{b}}^{\mathrm{i}} = \frac{\int_{\theta_A} P(\theta) \cdot T_{\mathrm{b}}^{\mathrm{i}} \cdot \mathrm{d}\theta}{\int_{\theta_A} P(\theta) \cdot \mathrm{d}\theta}, \qquad (4.5)$$

where

$$\frac{P(\theta)}{P(0)} = \left(\frac{J_1(x)}{x} \right)^2, \qquad x = \frac{\pi d}{\lambda} \sin \theta$$

J_1 is the one-order Bessel function of the first kind, and λ the wavelength.

To determine the significance of this technique for in-flight calibration of the radiometer, the calculations of angular brightness temperature distribution at frequencies of 22 GHz and 75 GHz are made (see Figure 4.6 and Figure 4.7).

These figures present the theoretical T_b angular distribution for both polarizations. It is significant that the maximum of T_b on vertical polarization, corresponding to the Brewster angle (see 1 in Figure 4.6 and Figure 4.7), disappears, when the 'polarization coupling' effect is taken into account (see 3 in Figure 4.6 and Figure 4.7). The antenna pattern smoothing results in displacement and decrease in the brightness temperature maximum for both vertical and horizontal polarization. This is purely the maximum which corresponds to the hot scene that can be used for radiometer calibration (Figure 4.5). As for the cold scene, it is strongly measured at a viewing angle more than 10° above the horizon.

The discrepancies of about 3–4 K, which occur between the measurements and calculations on determining the maximum and minimum of brightness temperatures, can be attributed to one or more of the following: failure to use the total antenna pattern, error in sky model profile, or error in using meteorological variables. However, the good agreement between the measured calculated results demonstrated that the above error sources are negligible.

In addition, the calibration procedure described of the airborne radiometer can be effectively used for the remote sensing of the ocean surface if the antenna cross-

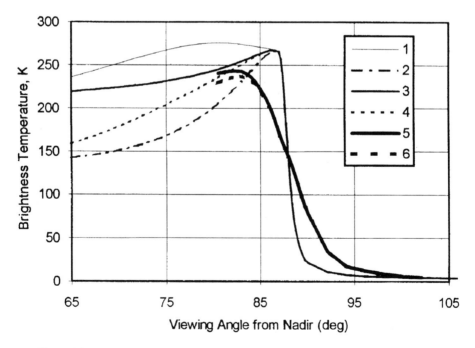

Figure 4.6. Theoretical brightness temperature angular distribution for 22 GHz in the vertical (1. 3, 5) and horizontal (2, 4, 6) polarization: 1, 2—T_b; 3, 4— T_b^i; 5, 6—$\tilde{T}_b^i$. Calculation for a flat sea surface, $d = 120$ mm, $\alpha = 25°$; other parameters correspond to Figure 4.5.

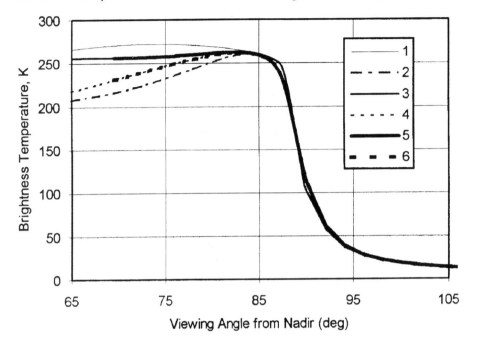

Figure 4.7. Same as Figure 4.6, but for 75 GHz.

polarization term is less than −22 dB. Otherwise, the polarization aspects of the antenna and observed scene should be taken into account [18, 19].

4.4 'METEOR-3M' MTVZA SATELLITE MICROWAVE RADIOMETER

From 1999 the Meteorological Satellite Program of the Russian Space Agency will use the advanced 'Meteor-3M' spacecraft [23]. The series of satellites with microwave instruments aboard will be launched in sun-synchronous orbit. The MTVZA multichannel microwave imaging radiometer will be used for remote sensing of the ocean and land surface, and integrated parameters of the atmosphere, as well as for measuring the global atmospheric temperature and water vapor profiles. In addition, the MTVZA includes some complementary non-typical operating frequencies especially for oceanographic research. The schedule of orbiting the 'Meteor-3M' spacecraft and parameters of the orbits are shown in Table 4.1.

The MTVZA satellite microwave radiometer is based on the block-diagram of the airborne multichannel imaging radiometer (see section 4.1). MTVZA operating frequencies are located in the transparent window of atmosphere 19, 33, 36.5, 42, 48, 91.65 GHz, in the absorbing lines of oxygen 52–57 GHz and water vapor 22.235 and 183.31 GHz. The instrument will provide measurements of the atmosphere temperature profile to approximately 40 km and water vapor profile to 6 km.

Table 4.1. Orbit parameters for 'Meteor-3M' spacecraft

Spacecraft	Launch date	Inclination (deg.)	Altitude (km)	Period (min)	Ascending equator crossing time (local)
'Meteor-3M', No. 1	Apr. 1999	99.64	1024	105.53	09.15
'Meteor-3M', No. 2	Aug. 2000	99.64	1024	105.53	10.30 (16.30)

All radiometer channels of MTVZA are switched to a single feed-horn antenna (Figure 4.8). The total-power radiometer configuration is employed which provides a factor of two greater sensitivity over a conventional 'Dicke' switched system. The expected receiver noise temperature is shown in Figure 4.8. The design and salient characteristics of this instrument are shown in Tables 4.2 and 4.3.

The antenna system of MTVZA consists of an offset parabolic reflector of dimensions 50×65 cm, illuminated by a broad-band, eleven-port feed-horn antenna through the flat mirror. The two-mirror antenna system configuration is due to the instrument deploying in the bottom part of the spacecraft (see Figure 4.9). The reflector, flat mirror and feed-horn antenna are mounted on a drum, containing the radiometers, digital data subsystem, power and signal transfer assembly, which rotates continuously about an axis parallel to the local spacecraft vertical. The power, commands, all data, timing and telemetry signals pass through slip-ring connectors to the rotating assembly.

For calibration, the hot and cold reference absorbers are used. These are mounted on the non-rotating part of the instrument and are positioned such that they pass between the feed-horn and the flat mirror, occulting the feed-horn once each scan. The temperature difference between the hot and cold targets is expected to be 60–70 K.

The MTVZA will rotate continuously about an axis parallel to the local spacecraft vertical with a period of 2.52 s during which the subsatellite point, moving at 6.35 km/s, travels 16 km. The instrument is a conical scanning device, and looks backward. The viewing angle is 51.3° and the incidence angle with respect to the Earth's surface is 65°. The sampling resolution is expected to be 16 km for channels of 91.6 GHz and 183 GHz, and 32 km for other channels in both the cross-track and along-track directions. The scan direction will be from left to the right when looking in the aft direction of the spacecraft, with the active scene measurements lying ±60° about the aft direction, resulting in a swath width of 2600 km. This will result in 12 hours of global coverage for one satellite. Only regions of about 300 km width will be missed near the equator, which will be covered after 24 hours. Plans are for two MTVZAs to be in orbit at the same time, which will increase the timeliness of the coverage.

The processing and archiving of MTVZA microwave data will be carried out at the Center for Program Studies, Russian Space Agency (Web Home Page **http://cpi,rssi,ru**).

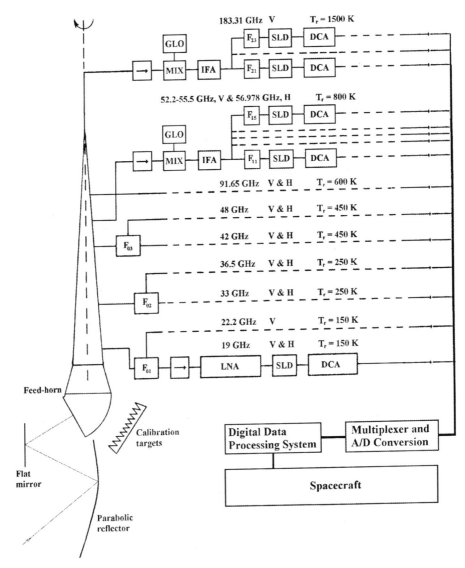

Figure 4.8. Block-diagram of MTVZA satellite microwave radiometer: F_{ij}—filter; MIX—mixer; GLO—Gunn local oscillator; IFA—intermediate frequency amplifier; LNA—low-noise amplifier; SLD—square law detector; DCA—direct current amplifier.

4.5 MICROWAVE RADIOMETERS OF THE DEFENSE METEOROLOGICAL SATELLITE PROGRAM (DMSP) SPACECRAFT

The Defense Meteorological Satellite Program USA has been in place since 1987. Its main goal is to obtain near-real-time global maps of cloud water; rain rates; water vapor over the ocean; marine wind speed; sea ice location, age, and concentration; snow water

Table 4.2. MTVZA performance characteristics

Frequency (GHz)	19	22.2	33	36.5	42.0	48	52–57	91.6	183
Polarization, V/H	V, H	V, H	V, H	V, H	V, H	V, H		V, H	
Spatial resolution (km)	75	68	45	41	36	32	30	18	12
Circular conical scanning period (s)					2.52				
Viewing angle (deg.)					51.3				
Incident angle (deg.)					65				
Swath width (km)					2600				
Mass (kg)					80				
Power consumed (W)					90				

Table 4.3. MTVZA frequency channel characteristics

Channel no.	Center frequency (GHz)	No. of pass bands	Band-width (MHz)	Approximate peak sensitivity altitude, km
1	19	1	1000	—
2	22.235	1	1000	—
3	33	1	2000	—
4	36.5	1	2000	—
5	42	1	2000	—
6	48	1	2000	—
7	52.28	1	400	2
8	52.85	1	300	4
9	53.33	1	300	6
10	54.40	1	400	10
11	55.45	1	400	14
12	56.968 ± 0.1	2	50	20
13	56.968 ± 0.05	2	20	25
14	56.968 ± 0.025	2	10	29
15	56.968 ± 0.01	2	5	35
16	56.968 ± 0.005	2	3	40
17	91.65	2	3000	Surface
18	183.31 ± 7.0	2	1500	1.5
20	183.31 ± 3.0	2	1000	2.9
21	183.31 ± 1.0	2	500	5.3

Figure 4.9. Russian Space Agency 'Meteor-3M' satellite with MTVZA microwave radiometer located bottom left.

content; land surface type, moisture, and temperature, as well as for measuring global atmospheric temperature and water vapor profiles. Three types of microwave radiometer are used for solving these problems. The Special Sensor Microwave/Imager (SSM/I) provides measurements of Earth surface parameters and meteorological variables of the atmosphere; the Special Sensor Microwave Temperature Sounder (SSM/T-1) provides the vertical temperature profile of the atmosphere; and the Special Sensor Microwave Water Vapor Profiler (SSM/T-2) provides the atmospheric water vapor profile [3, 20, 21].

The Global Hydrology and Resource Center (GHRC) of NASA Marshall Space Flight Center and the Global Hydrology and Climate Center have been processing and archiving SSM/I data for almost 5 years. Prior to May 1995, the SSM/I data source for the GHRC was the National Environmental Satellite Data and Information Service (NESDIS). Since May, 1995, the source has been the Fleet Numerical Meteorology and Oceanography Center (FNMOC). Data are obtained from FNMOC and processed at the GHRC within hours of receipt. Each day, full resolution or 'swath' brightness temperatures and reduced resolution 'gridded' data sets are generated. Browsed images of the gridded files are created in both HDF raster and GIF formats. HDF represents the Hierarchical Data Format, the data format standard for NASA's Earth Observing System Data and Information System. The GHRC is producing several geophysical products including: integrated water vapor, cloud liquid water, and oceanic wind speed. These products are in swath and gridded formats. (Web Home Page **http://ghrc.msfc.nasa.gov**).

The US Department of Defense maintains a series of polar orbiting sun-synchronous meteorological satellites as a part of the Defense Meteorological Satellite Program. To date, six of the DMSP satellites have flown with an SSM/I (see Table 4.4 and Table 4.5).

The SSM/I is a seven-channel, four-frequency, linearly polarized, passive microwave radiometric system [3]. It receives both vertically and horizontally polarized radiation at 19.3, 37.0, and 85.5 GHz and vertically only at 22.2 GHz. Table 4.6 shows the design and salient characteristics of this instrument.

Figure 4.10 shows the conical scan of the SSM/I in the early mooning orbit for DMSP F-8. The active portion of the scan covers a swath of 1400 km, which results in the 24-hour global coverage for one satellite. Only the diamond-shaped areas near the equator are missed. The small circular sectors of 280 km radius at the north and south poles are never measured owing to the orbit inclination, but the diamond-shaped regions

Table 4.4

DMSP spacecraft	Launch date	SSM/I instrument status
F-8	19 June 1987	Operational support ended 13 Aug 1991
F-10	1 Dec 1990	Operational support ended 14 Nov 1997
F-11	28 Nov 1991	Currently functioning
F-12	29 Aug 1994	Failed shortly after launch
F-13	24 Mar 1995	Currently functioning
F-14	4 Apr 1997	Currently functioning

Table 4.5 Orbit parameters for DMSP spacecraft

DMSP spacecraft	F-8	F-10	F-11	F-12	F-13	F-14
Launch date	19/07/87	01/12/90	28/11/91	29/08/94	24/03/95	04/04/97
Inclination (deg.)	98.8	98.8	98.8	98.9	98.8	98.8
Ascending equator crossing time (local)	06:15	22:25	18:25	21.27	17:43	20:29
Orbit period (min)	101.8	100.5	101.9	102.0	101.9	101.9
Max altitude (km)	882	841	878	851	856	
Min altitude (km)	838	728	841	844	844	

are covered after 72 hours. Because there are at least three or four SSM/Is in orbit at the same time, the timeliness of the coverage is greatly increased.

The SSM/I consists of an offset parabolic reflector of dimensions 61×66 cm, illuminated by a corrugated, broad-band, seven-port horn antenna. The reflector and feed-horn antenna are mounted on a drum which contains the radiometers, digital data subsystem, mechanical scanning subsystem, and power subsystem. The entire reflector, feed-horn, and drum assembly is rotated about the axis of the drum by a coaxially mounted bearing and power transfer assembly. The spin rate is 31.6 rev/min. All data, commands, timing and telemetry signals, and power pass through it on slip-ring connectors to the rotating assembly.

The SSM/I calibration system consists of a small mirror and a hot reference absorber, which are not rotated with the drum assembly. They are positioned off axis such that they pass between the feed-horn and the parabolic reflector, occulting the feed-horn once each scan. The mirror reflects cold 2.7 K cosmic background radiation into the feed-horn, thus serving, along with the hot reference absorber. This scheme provides an overall end-to-end absolute calibration which includes the feed-horn. The absolute brightness temperature calibration accuracy is better than ±3 K.

A total power radiometer configuration is employed which provides a factor-of-two greater sensitivity over a conventional 'Dicke' switched system. Both the calibration scheme and the use of total power radiometers are innovations, significantly improving the performance of the SSM/I compared to previous spaceborne radiometers.

The SSM/I rotates continuously about an axis parallel to the local spacecraft vertical with a period of 1.9 s during which the subsatellite point, moving at 6.58 km/s, travels 12.5 km. The scan direction is from left to right when looking in the aft direction of the spacecraft (see Figure 4.10), with the active scene measurements lying ±51.2° about the aft direction, resulting in a swath width of 1400 km. Beginning with the DMSP F-10 spacecraft, the SSM/I scanning configuration is in a forward direction. The viewing angle is 44.8° and the incidence angle with respect to the Earth surface is 53.1°.

The applications of SSM/I data for oceanographic research are discussed in section 5.8.

The Special Sensor Microwave Temperature Sounder (SSM/T-1) [20] is flown aboard the DMSP F-11, F-12, F-13, and F-14 satellites. The SSM/T-1 sounder consists of seven

Figure 4.10. Defense Meteorological Satellite Program (DMSP) Block 5D-2 satellite with the
Special Sensor Microwave/Imager (SSM/I) located at the upper left [3].

microwave channels with atmospheric sensitivity in the oxygen absorption band, at
frequencies ranging from 50.5 to 59.4 GHz, with one channel acting as a surface window
channel. Channel parameters and regions of peak sensitivity in the atmosphere are given
in Table 4.7.

Table 4.6. SSM/I performance characteristics

Channel frequency (GHz)	Polarization, V/H	IF passband (MHz)	Footprint on Earth's surface (km)		Footprint sampling resolution (km)
			Along-track	Cross-track	
19.35	V	10–250	69	43	25
19.35	H	10–250	69	43	25
22.235	V	10–250	60	40	25
37.0	V	100–1000	37	28	25
37.0	H	100–1000	37	29	25
85.5	V	100–1500	15	13	12.5
85.5	H	100–1500	15	13	12.5

Table 4.7. SSM/T-1 performance characteristics

Channel	Frequency (GHz)	Bandwidth (MHz)	Approximate peak sensitivity	
			Altitude (km)	Pressure (Mb)
1	50.50	400	Surface	—
2	53.20	400	1	800
3	54.35	400	6	400
4	54.90	400	10	280
5	58.40	115	30	10
6	58.825	400	18	70
7	59.40	250	23	40

The radiometer antenna completes one scan of seven Earth-view measurements and two calibration measurements every 32 s. In order to obtain the seven Earth-view measurements, the antenna is rotated in seven equal angular increments, resulting in a center nadir view and six limb views. At nadir, the antenna observes a circular area (footprint) of 175 km in diameter at the Earth's surface. As the antenna scans away from the nadir on either side, the footprint becomes elliptical, increasing to 230×305 km at the largest viewing angle. At each viewing angle, the instrument collects measurements for each of the seven channels. Each scan of the instrument covers a swath of 1600 km perpendicular to the orbital tract with 195 swaths per orbit.

The SSM/T-1 sounding products are disseminated in near real time by a variety of methods for use in local analyses and numerical weather prediction forecast models. The resulting sounds products are mean lay virtual temperatures at 30 atmospheric levels from the surface to 10 Mb.

The Special Sensor Microwave Water Vapor Profiler (SSM/T-2) [21] is flown aboard the DMSP F-11, F-12, F-13, and F-14 satellites. The SSM/T-2 is the first operational microwave water vapor sounder to be placed in orbit. It is a scanning, five-channel, passive and total power microwave radiometer system. The five channels consist of three water vapor channels centered around the 183.31 GHz water vapor line, a 91 GHz channel, and a 150 GHz channel. Table 4.8 lists the channel characteristics. The SSM/T-2 observation rate is 7.5 scans per minute. There are 28 observations (beam positions) per scan for each of the five channels, with each observation having a spatial resolution of approximately 48 km. All five channels have coincident centers. The total swath width for the SSM/T-2 is approximately 1400 km.

Table 4.8. SSM/T-2 performance characteristics

Channel	Frequency	IF bandwidth (GHz)	Nadir field-of-view (km)	Approximate peak sensitivity (Mb)
1	91.665	1.5	84	Surface
2	150.0	1.5	54	1000
3	183.3 ± 7	1.5	48	800
4	183.3 ± 3	1.0	48	650
5	183.3 ± 1	0.5	48	500

The SSM/T-2 employs a single offset parabolic reflector with a 2.6-inch diameter projected aperture. The reflector is shrouded to eliminate the possibility of rays from the Sun striking either of the calibration paths and causing unwanted thermal gradients. The feed-horn is a corrugated pyramidal horn with a flare designed to minimize phase center separation over the bandwidth (91 to 183.3 GHz), while providing a spherical wave illumination of the reflector. A 3.3° beamwidth is achieved for the 183.3 GHz channels and larger beamwidths of approximately 3.7° and 6.0° for 150 and 91.665 GHz, respectively. These correspond to the field-of-view (FOV) parameters given in Table 4.8.

SSM/T-2 provides global monitoring of the concentration of water vapor in the atmosphere under all sky conditions by taking advantage of the reduced sensitivity of the microwave region to cloud attenuation.

There are plans to use the new instrument SSM/IS, which will provide all of the products of the SSM/T plus those of the atmospheric temperature and water vapor sounders, SSM/T-1 and SSM/T-2. Five of these sensors will be flown on DMSP spacecraft F-16 through F-20 [22].

REFERENCES

[1] Ulaby, F.T., Moor, R.K., and Fung, A.K. (1981) *Microwave Remote Sensing: Active and Passive*. Vol. 1, New York.

[2] Njoku, E.G., Stacy, J.M., and Barath, F.T. (1980) 'The Seaset Scanning Multichannel Microwave Radiometer (SMMR): instrument description and performance', *IEEE J. Oceanic Engineering*, **OE-5**, No. 2, 100–115.

[3] Holinger, J.P., Pierce, J.L., and Poe, G.A. (1990) 'SSM/I Instrument and Evaluation', *IEEE Trans. Geosci. and Remote Sensing*. **28**, No. 5, 781–790.

[4] Armand, N.A. (1993) International Project PRIRODA: Instruments. Reference Handbook. Institute of Radioengineering and Electronics RAS, Moscow, September.

[5] Racette, P.E., Dod, L.R., Shiue, J.C., Adler, R.F., Jackson, D.M., Gasiewski, A.J., and Zacharias, D.D. (1992) 'Millimeter-wave imaging radiometer for cloud, precipitation, and atmospheric water vapor studies', in *Proceedings of IGARSS'92 Symposium*, Houston, TX, Vol. 2, pp. 1426–1428.

[6] Etkin, V.S., Aleksin, B.E., Aniskovich, V.M., *et al.* (1987) 'Airborne multichannel complex for radio-hydrophysical studies', IKI RAN Preprint (Space Research Institute) N1279, Moscow, p. 44.

[7] Wilson, W.J., Howard, R.J., Ibbot, A.C., Parks, G.S., and Ricketts, W.B. (1986) 'Millimeter-wave imaging sensor', *IEEE Trans. on Microwave Theory and Techniques*, **MTT-34**, No. 10, 1026–1035.

[8] Cherny, I.V. (1982) 'MM-wave radiometer–scatterometer for remote sensing of sea surface', IKI RAN Preprint (Space Research Institute) No. 689, Moscow, p. 19.

[9] Gorobetz, N.N., Zabyshny, A.I., Il'gasov, P.A., Cherny, I.V., and Sharapov, A.N. (1989) 'Multichannel microwave radiometer for remote sensing of ocean and atmosphere', IKI RAN Preprint (Space Research Institute) No. 1545, Moscow, p. 32.

[10] Cherny, I.V., Alesin, A.M., Gorobetz, N.N., Nakonechny, V.P., Pantzov, S.Yu., and Zabyshny, A.I. (1995) 'Advanced airborne multi-spectral mm-wave imaging technique for ocean and atmosphere studies', in Proceedings of CO-MEAS'95 Symposium, Atlanta, Georgia, April.

[11] Feldstein, A.L., Yavich, L.R., and Smirnov, V.P. (1967) *Waveguide Handbook.* "*Sov. Radio*", Moscow.

[12] Timofeeva, A.A. (1977) 'Determination of feed-horn parameters with the near optimal antenna pattern', *Russian J. "Electrosvyas"*, No. 5, pp. 28–33.

[13] Stepanenko, V.D., Schukin, G.G., Bobylev, L.P., and Matrosov, S.Yu. (1987) *Radio-Thermosensing in Meteorology.* Leningrad: Gidrometeoizdat.

[14] Liebe, H.J. (1989) 'MPM—an atmospheric mm-wave propagation model', *Int. J. Infrared Millimeter Waves*, **10**, No. 4, April.

[15] Pandey, P.C. and Kakar, R.K. (1982) 'An empirical microwave emissivity model for a foam covered sea', *IEEE J. Oceanic Engineering*, **OE-7**, No. 3, 135–140.

[16] McKinley, S.C. and Philbrick, C.R. (1993) 'Tropospheric water vapor concentration measured in Penn State/ARL lidar', in *Proceedings of CO-NMEAS'93 Symposium*, Albuquerque, NM, pp. 185–188, March.

[17] Gasiewski, A.J., and Kunkee, D.V. (1993) 'Calibration and applications of polarization-correlating radiometers', *IEEE Trans. Microwave Theory and Tech.*, **41**, No. 5, 767–772.

[18] Claassen, J.P. and Fung, A.K. (1974) 'The recovery of polarized apparent temperature distributions of flat scenes from antenna temperature measurements', *IEEE Trans. Antenna and Propagation*, **AP-22**, No. 3, 433–442.

[19] Beck, F.B. 'Antenna pattern corrections to microwave radiometer temperature calculations', *Radio Science*, **10**, No. 10, 839–845.

[20] Donahue, D., Pettey, M., Whistler, B., and Kratz, G. (1994) 'DMSP SSM/T-1 physical retrieval system', *Critical Design*, May.

[21] Falcone, V.J., Griffin, M.K., Isaacs, R.G., Pickle, J.D., Morrissey, J.F., Bussey, A., Kakar, R., Wang, J., and Racette, P. (1993) 'SSM/T-2 calibration data analyses', in *Proceedings of CO-MEAS'93 Symposium*, Albuquerque, NM, pp. 165–168, March.

[22] Holinger, J.B. (1990) 'Introduction', *IEEE Trans. Geosci. and Remote Sensing*, **28**, No. 5, 779–780.

[23] Cherny, I.V., Chernyavsky, G.M., Khapin, Yu B., and Volkov, A.M. (1997) Applications of 'Meteor-3M' satellite microwave radiometers for oceangraphic research and early diagnostics of natural disasters'. In *Proceedings of Fourth International Conference on Remote Sensing for Marine and Coastal Environments*, Orlando, Florida, Mach.

5

Microwave observations of processes in the ocean–atmosphere system

5.1 NEW APPROACH FOR MICROWAVE DIAGNOSTICS OF THE DEEP OCEANIC PROCESSES BASED ON THE AMPLIFICATION MECHANISM CONCEPTION

Passive microwave methods for the remote sensing of the ocean–atmosphere system are used widely for determining meteorological parameters of the atmosphere and ocean surface [1–7]. For example, the Special Sensor Microwave/Imager (SSM/I), operating at frequencies of 19.35, 22.235, 37.0 and 85.5 GHz, provides the operative standard meteorological data [8]. In general, the satellite passive microwave data are used to produce the many environmental products: ocean surface wind speed [9–11]; integrated water vapor [12]; cloud liquid water content [13]; precipitation; ice coverage and age; snow water content [14, 15]. At present, the approach is developed to determine both ocean surface wind speed and wind direction, using passive microwave data [11, 16].

It is possible to study some oceanic or atmospheric processes by measuring certain parameters [17, 18]. Nevertheless, the links between the ocean and atmosphere form a closed loop with no beginning and no end. Although some oceanic and atmospheric processes may be studied in isolation, we are acutely aware that true understanding will continue to elude us unless we study the oceans and atmosphere as a coupled system. This is especially true in the case of remote sensing of the oceans; since we are dealing almost exclusively with the properties of the ocean surface, we cannot avoid the influences of both the overlying atmosphere and the ocean active layer [19].

In fact, all the processes taking place on the ocean surface are related either directly or indirectly to the state of deep water layers, even if the surface phenomena occur under direct atmospheric influence. It is valid mostly for anomalous states of the surface.

Nowadays, an important problem in the remote sensing community is the ability to 'look into' the ocean depths through the surface. This statement of the problem is quite correct. Indeed, the ocean is a thermodynamically non-equilibrium medium due to the presence of currents, temperature and salinity gradients, which form the ocean thermohaline fine structure. Therefore, the favorable conditions could arise which amplify the influence of deep water processes on the surface at the expense of accumulated oceanic energy [20].

The development of shear current instability, for example, results in amplification of oceanic internal waves; thus the energy of the current is transferred to a wave, increasing its amplitude and, in consequence, amplifying the modulation of surface waves. The said processes are called the Kelvin–Helmholtz instability and emissivity instability [21, 22].

Similar conditions arise when the non-equilibrium sources are the temperature and salinity gradients. The existence of a thermohaline fine structure of the ocean promotes the development of different types of instability, in particular   convective instability, otherwise known as 'double diffusion' [23, 24], which can serve as an indicator of some processes in the ocean (currents, internal waves, eddies and Rossby waves, upwelling), and atmosphere (typhoons, hurricanes, heavy tropical rainfalls).

In this case, the signal registered by remote means at microwave frequencies is being formed under the direct influence of two factors. First, the process under investigation, taking place at the ocean depth. And, second, certainly connected with the first one and directly characterizing the state of the non-equilibrium oceanic medium. The diagnostics of this secondary process may, in some cases, result in more information than that of the first [20].

Let us consider this concept in more detail. From the thermodynamic point of view the evolution of a non-equilibrium system always obeys the following law—the quickest relaxation to an equilibrium, which is the greatest possible under the given conditions. One of the ways of such relaxation could be served the development of secondary instability, including the spontaneous infringement of symmetry in the system [25, 26]. From the linear stability analysis the preference is given back to modes having the greatest growth rate. However, in a number of cases the symmetry of a problem is those, that in taking into account the non-linear members in the equations the system evolution, is essentially slowed down in frameworks of the originally chosen decision from the linear analysis and for system is 'profitable' to break spontaneously initial symmetry, due to secondary instability, to create the conditions for fastest relaxation. We shall note, that is just the large-scale vortices in the atmosphere (tropical cyclones) that are generated as a consequence of energy swapping from small-scale turbulence to large-scale structures (inverse spiral cascade) [27], the ocean thermohaline fine structure is formed [23] etc. The above-mentioned examples show that the most characteristic way for system evolution due to the secondary instabilities is the developing of subsystems not being quasi similar to the basic one.

The special interest is represented by the phenomena, developing on a threshold of stability (or near it) for a system in which the secondary instability does not spontaneously arise. Such a condition is sufficient for many ocean regions, especially in energy-active zones with intensive currents, temperature and salinity gradients [23, 24, 28]. As a result of primary instability the non-linear evolution is sated, but the background currents, temperature and salinity gradients do not usually disappear, but only change, so that the system is on a stability threshold (i.e. system has a certain 'stock' of accumulated energy). It appears that if the wave signal of finite amplitude and certain structure gets into such a medium, its energy can be essentially increased [25, 26]. At that point, the energy of the medium is also changed. This is important for remote sensing methods because, in some cases, it can be easy to detect the changes in the medium's state. Figure 5.1 demonstrates the way for diagnostics of the deep oceanic processes and natural disasters with the ocean surface microwave emission based on the amplification mechanism conception.

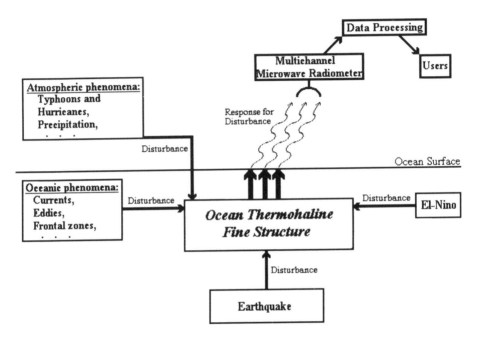

Figure 5.1. The concept for diagnostics of the deep oceanic processes and natural disasters with ocean surface microwave emission [20].

The ocean surface microwave phenomenon due to this secondary process is characterized by spectral and polarization variations in brightness temperature and is quite different from those caused by the wind stress and direct current influence onto the surface, as well as meteorological variables of the atmosphere [20]. The notion of interchannel derivative is used for the analysis of the multichannel radiometer signals to separate the oceanic processes from the atmospheric ones. Using the said classification sign, we studied many processes in the ocean–atmosphere system from the small scale up to synoptic ones which are accompanied by the spectral and polarization variations in ocean surface brightness temperature.

The amplification model has been confirmed experimentally and can form the basis of the development of new methods and instruments for remote sensing of the ocean–atmosphere system. For effective diagnostics of the secondary non-stable processes, microwave radiometers have to include a number of non-typical remote sensing frequencies. Below it is illustrated by many field experiments, which authors personally carried out by means of microwave radiometers deployed on ships and aircraft, as well as processing the data from the SSM/I satellite microwave radiometer. The said ocean surface microwave phenomenon caused by the oceanic processes was detected during different weather conditions and surface states, when the wind speed was from 3–4 m/s up to 20–25 m/s.

The method for remote diagnostics of ocean–atmosphere system state has been patented [29, 30].

5.2 OBSERVATIONS OF OCEANIC INTERNAL WAVES AND SOLITONS

Remote observations have grown into an important source of information on internal waves in the ocean [31–37]. Only from such observations can we obtain data on the broad horizontal structure of the internal waves. The basic trend in research is a detailed comparison of remote measurement results with *in situ* measurements of the oceanographic parameters, with the object of determining the mechanisms by which internal waves act on the ocean surface.

Shipboard microwave observations of internal waves

Here, we describe certain results from simultaneous observations of the manifestation of strong internal waves at the surface by means of shipboard microwave instruments and contact-transducer measurements of such waves in the Indian Ocean during the 26th cruise of R/V *Dmitry Mendeleyev* and solitary internal waves in the Sea of Japan during the 11th cruise of R/V *Professor Bogorov*.

One of the basic results of the R/V *Dmitry Mendeleyev* expedition was the detection of a broad field of strong high-frequency internal waves in the region of the Mascarene Plateau. The test-range measurements made here indicated that particularly strong internal waves were generated by tidal currents on the eastern slopes of the ridge and propagated eastward into a region of typical ocean depth. While the ship was in this part of the open ocean, contact transducers repeatedly recorded trains of internal waves that often reached amplitudes of tens of meters. The manifestation of the internal waves on the ocean surface was classical, in the form of stormy bands of breaking surface waves and comparatively smooth bands in which the waves were very light. The bands alternated in space at intervals of 0.5–2.5 km and had a temporal period of ~ 10–20 min, i.e. they propagated across the surface with velocities in the 0.5–2 m/s range. The sharp periodic changes in the physical state of the air–sea interface were easily recorded by remote microwave means. The measurements were made with the ship both under way and crossing several bands of alternating surface waves, i.e. internal-wave packets, and from the drifting ship, in which case the packet would move slowly across beneath the ship. The pattern of shipboard experimental measurements of internal waves by remote means is shown in Figure 5.2.

Microwave observations of the ocean surface were carried out by means of a radiometer–scatterometer operating at a frequency of 37 GHz with fluctuation sensitivities of ⩽ 0.2 K and > 60 dB in passive and active mode [38] and a radiometer operating at a frequency of 1.7 GHz with a sensitivity of ⩽ 0.2 K (the data on the sensitivities in the radiometer channels are given for a time constant of 1 s). Microwave instruments were deployed on a gyroscopically stabilized turntable 11 m above the water line. The viewing angle was 75°.

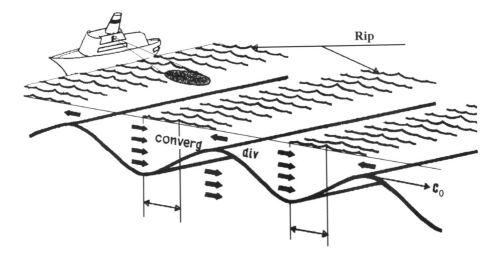

Figure 5.2. Pattern of shipboard experiment on microwave observations of internal oceanic waves.

An isotherm time curve recorded by a distributed temperature sensor lowered to a depth of 125 m is presented here as a contact-measurement result. The operating principle of the sensor was described in [39]. However, we note that this device basically registers the lower mode of the thermocline oscillations. A CTD-sonde was also used for standard hydrological measurements. An 'Elac' acoustic echo sounder was used in some of the measurements when sound-scattering layers were present.

The typical undisturbed hydrology of the upper layer of the ocean in the region in which the internal waves were observed ($12°23.7'$S, $61°44.4'$E) at 15.30 on February 9, 1981, is represented in Figure 5.3(a) by CTD-sonde vertical profiles. The presence of a thoroughly warmed subsurface layer much less than 200 m thick with a comparatively low density is characteristic here. The ocean was therefore stratified only in a narrow range of depth compared to its local depth (2860 m at the point at which the profiles in Figure 5.3 were recorded), and constituted a weakly non-linear dispersion medium for long internal waves. Steady-state disturbances in such a medium [40] take the form of non-linear, including solitary, waves. The profile of the Brunt-Visalia frequency N given in Figure 5.3(b) for the same conditions indicates that the core of the pycnocline was at depths of 50 to 120 m.

A sensor that averaged the temperature readings in the layer between the 120 and 130 m levels was placed near this core. Figure 5.4(a) shows a record of the temperature–sensor signal obtained on February 10, 1981, as the ship drifted near $13°05.4'$S, $61°58.9'$E, between 06.04 and 07.52 of local solar time. The average temperature readings were converted to displacements of the fundamental mode of the thermocline with the aid of an undisturbed water temperature profile $T°C$ recorded on the day before the arrival of the internal-wave train (Figure 5.3(a)). The largest displacement of the thermocline in the packet recorded here reached a total amplitude of ~ 40 m between 06.48 and 06.55. At these amplitudes, which are comparable to the thickness of the stratified layer, the internal waves did indeed become non-linear. Non-linear effects can be observed in

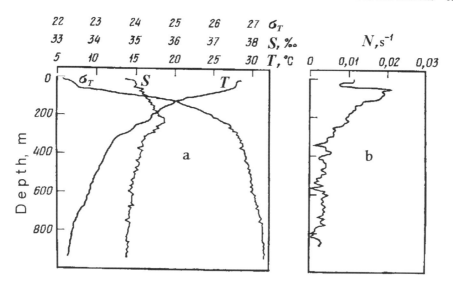

Figure 5.3. Vertical profiles of relative density σ_T, salinity S, temperature T (a), and Brunt-Visalia frequency N (b) from results of sounding at 15.30, on February 9, 1981, at 17°23.7'S, 61°44.4'E.

Figure 5.4(a) in the form of sharper notching of the troughs near the crest with the largest displacement [40] and asymmetry of the right-hand and left-hand slopes of the waves [41]. Nevertheless, the internal wave regime was far from forming solitary waves, an event that can occur on disturbances of similar intensity in a more sharply stratified upper ocean layer [40–42].

The results of remote measurements recorded in synchronism with the record of Figure 5.4(a) appear in Figures 5.4(b)–(d). The highest gradients of the measured geophysical parameters ranged up to very large values. The surface brightness temperatures varied with total amplitudes up to ~ 18 K at 37 GHz in the vertical polarization (Figure 5.4(c)) and up to ~ 3 K in the horizontal polarization at 1.7 GHz (Figure 5.4(b)), while the level of the backscattered vertical polarization signals at 37 GHz (Figure 5.4(d)) varied by ~ 10 dB. Here, the variations in ocean surface microwave emission at a frequency of 37 GHz (Figure 5.4(c)) were in phase with the thermocline oscillations (increasing as the latter approached the surface of the ocean), while the readings in the other two channels of the scatterometer and 1.7 GHz radiometer varied in anti-phase with the internal wave. The regular phase shift of the remote sensor signal oscillations with respect to the contact-transducer signal could be explained by the interaction of surface waves and internal waves. According to [43–46], the internal wave orbital velocities sweep the surface wave energy from regions of flow divergence and accumulate it in regions of flow convergence as shown in Figure 5.2. The convergence regions are just responsible for the rippled water and, in consequence, for the increase in scatterometer signal. In all cases, the phase shift is equal to a quarter of the internal wave period (Figure 5.4).

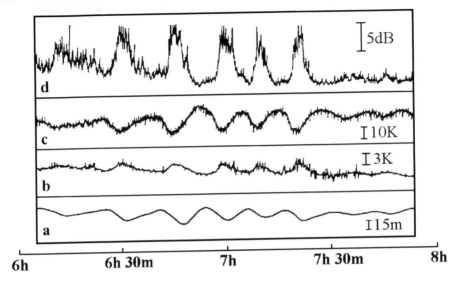

6h   6h 30m   7h   7h 30m   8h

Figure 5.4. Record of internal-wave train obtained between 06.04 and 07.52 on February 10, 1981, near 13°05.4′S, 61°58.9′E: (a) contact temperature sensor distributed at depths of 120–130 m; (b) 1.7 GHz radiometer with horizontal receiving polarization; (c) 37 GHz radiometer with vertical polarization; (d) 37 GHz scatterometer with vertical radiating and receiving polarization.

The almost exact reproduction of the internal-wave profile by the radiometer at 37 GHz is most impressive. This observational fact could not have been predicted in view of the 125 m layer of ocean separating the extremely thin skin layer on the surface, in which this shortwave radiation is formed, and the thick, deeper waters with the highest density gradient, in which the temperature sensor was deployed. We note that the largest cross-correlation coefficient between the distribution of brightness temperature at 37 GHz (Figure 5.4(c)) and the trend of the isotherm (Figure 5.4(a)) reached 0.8 even at the limited (no more than 100 elements) sample volume. Figure 5.5 shows the regression dependence between the increment of ocean surface brightness temperature at a frequency of 37 GHz (vertical polarization) and internal wave amplitude. From this it follows that the linear regression coefficient equals 0.6 K/m. In general, the quantitative relation between an oscillation amplitude of thermocline and increment of ocean surface brightness temperature could be changed and depends upon the depth of thermocline, wind speed and mutual propagation directions of internal and surface waves.

The stability of this picture is confirmed by extended-term microwave observations of internal waves that have been made on this test range. For example, Figure 5.6 shows a 7 hour internal wave record obtained as the ship drifted on a straight-line path between 20.00 on February 10, 1981 (13°01.2′S, 61°48.6′E) and 03.00 on February 11, 1981 (12°55.8′S, 61°38.1′E), the same channels as the record of Figure 5.4. The diagram in Figure 5.6(a), which reflects the position of the main sound-scattering layers at depths of 40–80 m, in supplementary here. The records in the other channels were graduated in accordance with the record in Figure 5.4. We see that the position of the sound-scattering layer (Figure 5.6(a)) practically coincides with that of the isotherm in the core of the

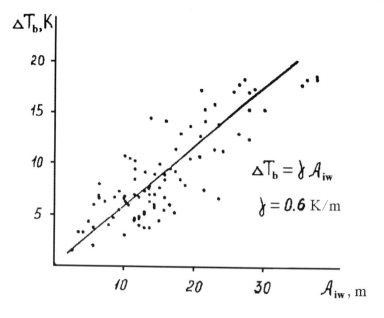

Figure 5.5. Regression dependence between sea surface brightness temperature variations at a frequency of 37 GHz (vertical polarization) and amplitude of internal waves.

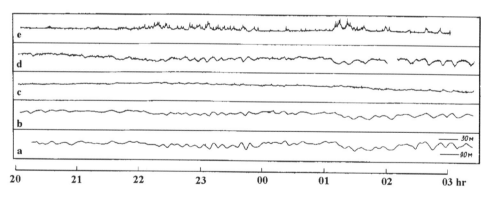

Figure 5.6. Seven-hour record obtained on February 10–11, 1981, from ship drifting between 13°01.2′S, 61°48.6′E and 12°55.8′S, 61°38.1′E: (a) 'Elac' echo depth sounder (30 and 90 m horizons are indicated on the diagram); (b) contact temperature sensors; (c) 1.7 GHz radiometer in horizontal polarization; (d) 37 GHz radiometer in vertical polarization; (e) 37 GHz scatterometer in vertical polarization.

pycnocline (Figure 5.6(b)), which is evidence of an exclusively one-modal nature of the internal waves. Even over 7 hours of measurements, the brightness temperature of the ocean surface in the 37 GHz channel (Figure 5.6(d)) reproduces the displacement in the internal wave with very high accuracy. The 1.7 GHz radiometer 'tracks' the internal waves in the same way (Figure 5.6(c)), but with stronger distortion of shape and lower sensitivity as compared with 37 GHz. The scatterometer reproduces internal waves with high amplitudes, greater than 10 m, quite reliably, but is practically insensitive to waves with amplitudes below 3 m, i.e. it has, for internal waves, a typical non-linear 'transfer' characteristic with amplitude cutoff.

Reliable registration of the surface manifestations of internal waves by the microwave instruments made it possible to measure the kinematic parameters of the waves. With the ship under way, it was possible, for example, to determine distances between individual internal waves and estimate their velocities of motion. The lower diagram in Figure 5.7 is a record obtained from the 1.7 GHz radiometer at ~ 19.00 on February 5, 1981, with the ship moving at a (through) speed of 8.5 knots. The course was set across the fronts of the internal waves, in the direction of their displacement. Three intersections of zones with intensified wave caused by internal waves were registered in the deviations of brightness temperature (~ 3 K from mean level) on this tack; they are represented by the open circles in Figure 5.7. Then, beginning at 20.00, 12°02.2′S, 61°14.2′E, the vessel began to back-track and, moving at 2 knots into the internal waves, the same radiometer recorded crossings of the zones noted earlier in reverse order (Figure 5.7, upper curve).

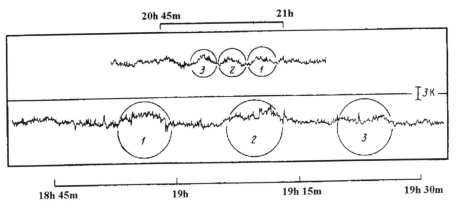

Figure 5.7. Record of packet of three internal waves obtained on February 5, 1981, by 1.7 GHz radiometer with ship under way near 12°23.2′S, 61°14.2′E (the internal waves are marked with numbered circles): lower curve—ship moves after packet, speed 8.5 knots; upper curve—ship moves toward packet, 2 knots.

The intervals between the internal waves on the following pack were around 15 min, and those on the opposing pack around 3 min. The differences between the internal waves can easily be calculated by adding the velocities of the ship on the two packs and dividing the sum by the sum of the reciprocal repetition intervals of the waves. For the conditions corresponding to Figure 5.7, this distance is approximately 800 m. On the other hand, the phase velocity of the internal waves relative to the ship was found to be quite high, just above 3 m/s. Considering the surface-current velocity (in this case 2.5 knots), the value $c_0 = 2$ m/s should be taken for the phase velocity.

The polarization of the radiometer–scatterometer was switched in some of the measurements. Figure 5.8 shows a record of a strong train of internal waves (amplitude of 60 m reached) that was registered from the drifting ship at ~ 18.30 on February 7, 1981, near 12°31′S, 59°40′E with the 37 GHz radiometer (Figure 5.8(b)) and scatterometer

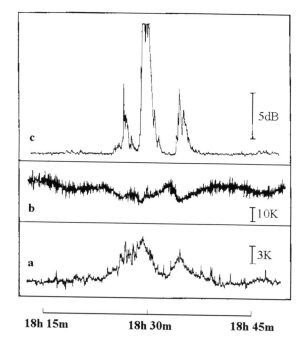

Figure 5.8. Horizontal-polarization signals received on February 7, 1981, near 12°31'S, 59°40'E:
(a) 1.7 GHz radiometer; (b) 37 GHz radiometer; (c) 37 GHz scatterometer.

(Figure 5.8(c)) channels in horizontal polarization. The record from the 1.7 GHz radiometer is included here as a reference signal for comparison. We can see from the figure that the manner in which the internal waves are indicated at 37 GHz depends significantly on polarization—the backscattering cross-section exceeded 15 dB, while the brightness temperature contrast of the ocean surface was only 10 K, although the internal waves amplitude in this experiment exceeds one as shown in Figure 5.4. The polarization of the 1.7 GHz radiometer was not switched at any time during the internal-wave studies.

To explain fully the remote-recorded characteristics of internal-wave surface manifestation, it is necessary first of all to invoke the concept of a multiple-scale non-linear hydrodynamic theory of the type developed in [40] in order to reconstruct the surface-velocity field from the available profile of relative density σ_T or the Brunt–Visalia frequency N (Figure 5.3) and the distribution of the isotherm (Figure 5.4(a)). It may be necessary to use numerical methods to obtain high computing accuracy. Then, considering the actual meteorological conditions of the observations, we can calculate, at least linearly, the modulation of the surface-wave spectrum by the currents due to the internal waves, as was done in [33]. Finally, by using electrodynamic empirical models, it is possible in principle to estimate, within the required ranges, the self-radiation and back-scattering gradients due to variations of the physical state of the ocean surface. Implementation of this program, as a whole and even in part, is a difficult task, and we shall therefore confine ourselves here to a qualitative description of the basic features of these remote-sensing results.

According to the contact-measurement results in Figure 5.4(a), the position of the isotherm during passage of the internal waves varied non-symmetrically by the undisturbed level at a depth of 125 m. The largest relative changes in the level of the isotherm ranged up to ~ 20% in the trough and ~ 10% on the crest of the strongest wave. Estimates made in the style of [40] for approximately equal thickness of the mixed layer and the intermediate stratified layer of the ocean (which corresponds to the diagram of Figure 5.3(b)) and for an internal-wave amplitude amounting to half this thickness gave rather high maximum velocities for the surface current moving with the wave: $U_m^- \approx 0.06c_0$ and especially for the counter-current, $U_m^- \approx 0.12c_0$. If $c_0 = 0.5$–2 m/s, these velocities amount to $U_m^+ = 3$–12 cm/s and $U_m^- = 5$–25 cm/s.

Surface gravity waves with group velocity c_0 should be captured by an opposing current with a very high velocity [33], i.e. they should grow sharply owing to synchronous pumping of energy from the internal wave and then break up. Typical length of these waves $\Lambda = 8\pi g^{-1}c_0^2$, where g is the acceleration due to gravity, lies in the meter range. It is wave breaking of this type that forms the rippled water ('rip') observed on the ocean surface in this experiment—the source of the high emission and backscattering gradients recorded by the remote sensing microwave instruments.

Internal soliton

It is significant to understand the interaction of internal solitary waves with the surface waves. This phenomenon has been studied in the Sea of Japan by means of a shipboard microwave radiometer at a frequency of 37 GHz [38] and a high-resolution echo depth sounder 'Elac'. The measurements were carried out during the 11th cruise of the research vessel *Professor Bogorov* on 18 October 1981 near the point of 42°37′N and 134°52′E. In this region the total original internal solitons are propagated in a north–south direction from the continental shelf to the deep water.

A microwave instrument was deployed at a height of 10 m from sea level. Remote sensing of the ocean surface was made on a vertical polarization at a viewing angle of 75°.

The typical undisturbed hydrology of the upper layer of the ocean in the region in which the solitary internal wave was observed is represented in Figure 5.9 by CTD-sonde vertical profiles. The sea depth was 2100 m.

Figure 5.10 shows the simultaneous measurements of the internal solitary wave at the depth and its manifestation on the sea surface. The ship's speed was 5 knots and it moved from south to north. Echo-printing (see Figure 5.10(a)) demonstrates the upper water layer within a depth range of 20–70 m. The undisturbed thermocline is disposed at a depth of 30–35 m. An internal soliton is a wave of depression, i.e. the wave peak is directed downward, and its amplitude is equal to 15 m.

The variations of sea surface brightness temperature (see Figure 5.10(b)) demonstrate the influence of the internal solitary wave on the surface. It should be noted that the shape of sea surface microwave emission variations do not repeat the profile of the internal solitary wave similar to those shown in Figure 5.4. We can see both an increase and a decrease in sea surface brightness temperature level due to internal soliton with respect to the background. The total brightness temperature contrast reaches about 8–9 K and to be comparable with one obtained from regression dependence at a frequency of

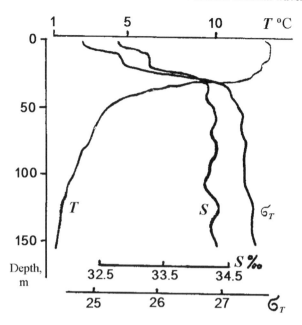

Figure 5.9. Vertical profiles of relative density σ_T, salinity S, and temperature T, from results of sounding at 21.20 on October 17, 1981, at 42°37′N, 134°52′E.

37 GHz, if an internal wave height is 15 m (see Figure 5.5). The sea surface wind speed was 4–5 m/s.

There are two disturbed areas of surface waves around the internal soliton. The rippled water area is located in front of the moving soliton, where brightness temperature level is lower. The slick area is characterized by increase in brightness temperature level and located behind the soliton. The above-mentioned features indicate that microwave observations of internal solitary wave allow us to define the direction of internal soliton propagation.

A model differential equation describing solitary wave behavior was not developed until 1895, when Kortweg and de-Vries approximated the Navier–Stokes equations for small finite-amplitude waves in a shallow channel.

In a two-layer fluid, the dimensional form of the Kortweg–de-Vries (K–dV) equation is given [42] by

$$\eta_t + c_0\eta_x + \alpha\eta\eta_x + \gamma\eta_{xxx} = 0. \tag{5.1}$$

Here, $\eta(x, t)$ is the interface displacement between the two fluids as a function of horizontal displacement x down the channel and at time t (see Figure 5.11), and c_0 is the phase speed of the associated linear wave. The subscripts in (5.1) refer to partial derivatives with respect to x or t. The upper layer is assumed to have depth h_1 and density ρ_1; in the lower layer the respective quantities are h_2 and ρ_2. For the constant coefficients of (5.1), we have approximately

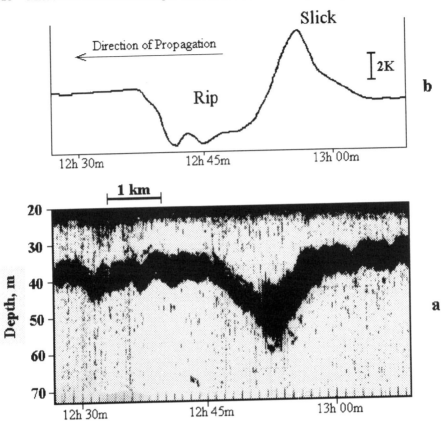

Figure 5.10. Record of internal soliton in the Sea of Japan on October 18, 1981, with ship under way near 42°37′N, 134°52′E: (a) echo depth sounder 'Elac'; (b) 37 GHz radiometer, vertical polarization (the integration time of the radiometer signal was 75 s).

$$c_0 \approx \left[g\left(\Delta\rho/\rho\right)h_1/(1+r)\right]^{1/2} \tag{5.2}$$

$$\alpha \approx -\left(3c_0/2\right)\left[(1-r)/h_1\right] \tag{5.3}$$

$$\gamma \approx c_0 h_1 h_2/6. \tag{5.4}$$

These approximate expressions were obtained by assuming $\rho \approx \rho_1 \approx \rho_2$, which is true in the ocean, where the small density differences are due primarily to temperature and salinity variations. Here, $\Delta\rho = \rho_2 - \rho_1$ and $r = h_1/h_2$.

The key to understanding this equation lies in the competition between the non-linear term ($\alpha\eta\eta_x$) and the dispersive term ($\gamma\eta_{xxx}$). Under certain conditions these terms balance, and the result is a stable configuration called the solitary wave, which is a special solution of (5.1) and has the analytical form

$$\eta(x,t) = -\eta_0 \text{sech}^2\left[(x-ct)/L\right], \tag{5.5}$$

where L is the characteristic length of the solitary wave, and we have assumed that the upper layer is thinner than the lower layer. This results in a downward displacement of the fluid interface, as indicated by the minus sign before η_0 in (5.5). The soliton phase speed is

$$c = c_0\left(1 - \eta_0\alpha/3c_0\right) \tag{5.6}$$

and the characteristic scale length is

$$L\left(-12\gamma/\eta_0\alpha\right)^{1/2}. \tag{5.7}$$

In our case, $\Delta\rho/\rho = 10^{-2}$, $\eta_0 = -15$ m, $h_1 = 35$ m, $h_2 = 2100$ m (see Figures 5.9 and 5.10), and in accordance with (5.6) and (5.7) we have $c = 1.45$ m/s and $L = 472$ m, respectively. The estimation of scale length in accordance with (5.7) is in good agreement with experimental data.

In our experiment the sea surface manifestation of the solitary wave is accompanied by rippled water, moving in the front of the soliton similar to those observed in the Andaman Sea [42] and shown in Figure 5.11. The rip on the surface is detected by the microwave radiometer and is related to a decrease in sea surface brightness temperature. But, in addition, the microwave radiometer registers the increase in sea surface brightness temperature around the internal solitary wave, which corresponds to the slick area on the surface. According to [47], the small surface soliton accompanies the internal solitary wave (see Figure 5.11). The analysis of microwave observations data and *in situ* measurements shows that it is just the slick on the surface that corresponds to the position of the surface soliton (see Figure 5.10).

Airborne microwave radiometric and radar observations of internal waves

New experimental results, associated with surface expressions of oceanic internal waves, were obtained during the Joint U.S./Russia Internal Wave Remote Sensing Experiment (JUSREX, 1992). Objectives, methodology, and some original data have been published in [48–50]. The experiment was conducted in the test area of the eastern end of Long Island, Atlantic Ocean. The experiment was organized principally by Johns Hopkins University Applied Physics Laboratory (JHU/APL) and the Space Research Institute, Moscow (IKI).

From previous remote sensing experiments (SARSEX, 1984) it was known that in the summer season the strong solitary internal waves are generated in the stratified regions of the continental shelf. Internal waves were observed systematically by synthetic aperture radars (SARs) as adjacent bright and dark bands on radar images. According to descriptions [48], the water column in this area has three distinct layers: a thin mixed layer from the surface to a depth of about 10 m; a strongly stratified region from 10 to 25 m depth; and a weakly stratified lower layer extending to the bottom. In the JUSREX experiment, the peak of the Brunt–Visalia frequency was about 30 cph and was located at a depth of 16–18 m. Typical internal wave propagation speeds were 0.6–0.7 m/s. Wavelengths of internal waves were 100–250 m, with wave crests separated by several hundred meters. Surface currents associated with internal waves were about 10–20 cm/s. The sea surface conditions were generally moderate.

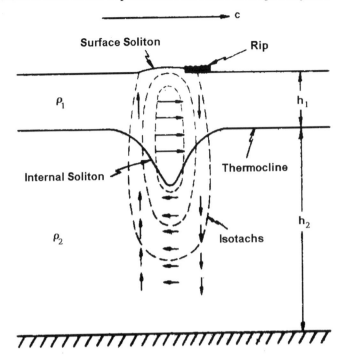

Figure 5.11. An internal soliton in a two-layer fluid of finite depth is a wave of depression when $h_1 < h_2$. Dashed lines are lines of constant water particle speed (isotachs); the arrows indicate the magnitude and direction of the water particles. The approximate location of the surface rips observed in the Andaman Sea by Osborne and Burch [42] is also shown. A small surface soliton accompanies the internal soliton [47].

Remote sensing systems included: satellite (ERS-1, 'Almaz') SAR, airborne (Tu-134, P-3, DC-8) Ku, X, C, and L-band radars, optical cameras, and multifrequency microwave radiometers and scatterometers. Shipboard oceanographic and meteorological instruments were used to measure *in situ* parameters of oceanic internal waves and surface currents simultaneously with remote sensing observations. Huge bodies of experimental data were collected during JUSREX; however, we call attention to the correlation features of radar and microwave radiometric images, illustrated spatial geometry, and signatures of internal waves.

JUSREX was the first to establish both microwave radiometric and radar signatures of intense packets of internal waves and solitons in the ocean. The spatial characteristics and scales of internal waves field were investigated during three levels of observations. The experiment also presented pioneer investigations of ambient hydrodynamic–electromagnetic processes, associated with internal wave–surface wave interaction, dynamics of surface currents, and influence of stable/unstable atmosphere conditions.

The Tu-134 aircraft carried a real-aperture side-looking airborne radar (SLAR), multifrequency microwave radiometers, and a six-band aerial photo camera. The corresponding specifications are presented in Tables 5.1 and 5.2. Figures 5.12 and 5.13 illustrate the aircraft laboratory and SLAR imaging geometry. Alternate H and V polarization radio impulses were transmitted from two suspension antennas, located under the fuselage. Backscattering signals of the same polarization were received simultaneously and produced four separate radar images (HH, VV, HV and VH polarizations). The SLAR swath was about 13 km on each side of the aircraft at an altitude of 2 km. Flight legs were usually 50–70 km.

Table 5.1. Tu-134 Ku-band SLAR specifications

Parameter	Value
Operating frequency	13.3 GHz ($\lambda = 2.25$ cm)
Transmitted power (peak)	60 kW
Transmitted pulse width	110 ns
Receiver bandwidth	16 MHz
Receiver sensitivity	−99 dB
Antenna beamwidth (azimuth)	0.0035 rad
Antenna dimensions	0.44×6 m
Swath width	12.5 km ($H = 2$ km)
Average geometric resolution	25 m × 25 m
Pulse repetition frequency	2 kHz
Polarization	VV, HH
Aircraft velocity	100–160 m/s
No. of integrated samples/pixel	180, nominal; function of velocity
Sampling rate	6 MHz × 8 bits
No. of pixels/row	512/512

Table 5.2. Tu-134 aircraft microwave radiometer specifications

Instrument	Frequency (GHz)	Wavelength (cm)	Δf (MHz)	ΔT (K) $\tau = 1$ s	Antenna beamwidth
R-18	1.6	18.6	125	0.10	30°
R-8	3.9	8.0	210	0.07	15°
RP-15 (3-channel polarimeter)	20.0	1.5	2000	0.15	9°
RP-0.8 (3-channel polarimeter)	37.0	0.8	1600	0.15	9°

The most important aspect of the SLAR imagery was the extreme sensitivity of the VV-polarization signal at the atmospheric boundary layer stability (see radar images in [48, 49]). Under stable atmospheric conditions, when the air temperature is higher than the surface water temperature, SLAR images on the vertical and horizontal polarization are qualitatively quasi-similar. The radar signature features are also similar, although the

Figure 5.12. The aircraft-laboratory Tupolev-134 with Ku-band SLAR, multifrequency microwave radiometers, and the MKF-6 six-band aero-photo camera.

contrast due to internal waves on the VV-polarization image is generally less than the
contrast on the HH-polarization image. Another picture is observed in the case of
unstable atmospheric conditions. The HH-polarization image at the top shows distinct
internal wave signatures. In the VV-polarization image a cellular-type structure masks the
internal wave signatures in the lower part of the image. The spatial scale of the structure
is a few kilometers. At the HH-polarization image, the internal wave signatures are again
presented.

Surface modulations induced by internal waves were measured *in situ* with the use of
radiometers, scatterometers, and radars installed on the bow of the R/V *Academician
Ioffe*. In particular, the period of brightness temperature variations correlated generally
with the main features of the radar internal wave signatures. The effect of atmospheric
stability on the brightness temperature variations was also manifest. It is important to
note that there is a spectral dependency of the sensitivity of microwave emission on
ocean–atmosphere conditions. During JUSREX, a high sensitivity of the R/V's radiometers to the parameters of internal waves was tested and was confirmed one more time.

Another situation occurred with the aircraft's radiometers. A set of radiometric data
was used primarily for reconstruction of the sea surface temperature (SST) and the wind
speed vector (WSV). By using the principle of 'polarization anisotropy' and semi-
empirical regression algorithms, one-dimensional distributions of SST and WSV along
each flight leg were designed. No modulation of radiometric signals due to the internal
wave's effect was found. Only in the case of large-scale cellular features induced by

a)

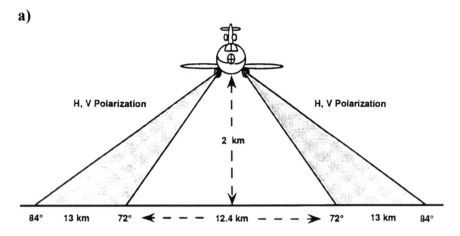

b)

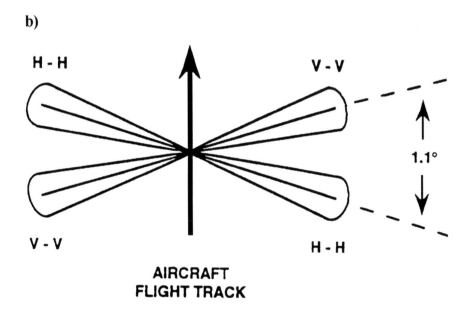

Figure 5.13. Diagrams of SLAR imagery (a) and antenna squint (b) [48].

atmospheric instability were modulations of radiometric signals observed. It was clearly an inconsistency between the spatial size of the surface roughness features induced by internal waves, and spatial resolution of the aircraft's radiometers (the spatial resolution is a few kilometers). At this averaging, only a large-scale change in the integral character-istics of the ocean surface conditions can be measured by microwave methods. In this case, no local effects, induced by a single internal wave, can be manifest, as opposed to the shipborne's microwave observations.

An alternative method of radiometric data processing was made by using a special two-dimensional statistical algorithm. The processing was applied to the spatial reconstruction of full microwave radiometric images of the test area where packets of internal waves were observed on the radar images. The main tool of computer processing lies in choosing a two-dimensional low-frequency filter (or smoothing window) for the determination of large-scale features on a microwave image associated with internal waves. Such a procedure was realized for the well-provided set of radiometric data.

Figure 5.14 shows the reconstructed radar image mosaic [50], and Figure 5.15 shows radiometric microwave images of the same test area with internal waves. For reconstruction of the microwave images, the radiometric data from three channels were used: $\lambda = 0.8$, 1.5 and 8 cm (at nadir view angle). Aircraft track reconstruction and radar mapping were made using navigation information according to five rosette legs (Figure 5.14, right-hand side). The radar image (HH polarization) contains a set of signatures from internal waves with different orientation, intensity and wavelength. This example reflects a great variety of oceanic internal wave patterns in the test area at moderate summer ocean conditions.

However, the passive microwave signatures of the same area with internal waves have a spatially distributed 'spotted' structure. On the microwave images (Figure 5.15), the brightest fragments indicate the most intensive non-linear internal waves or solitons, and correlate with corresponding radar signatures (Figure 5.14). The brightness contrast depends on the size of the smoothing filter. But qualitatively, the spotted-picture is maintained when the filter size is varied. Because the describing phenomenon has an

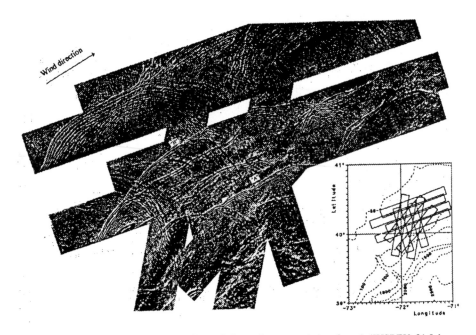

Figure 5.14. SLAR image mosaic of oceanic internal waves and aircraft track (JUSREX, 21 July 1992) [50].

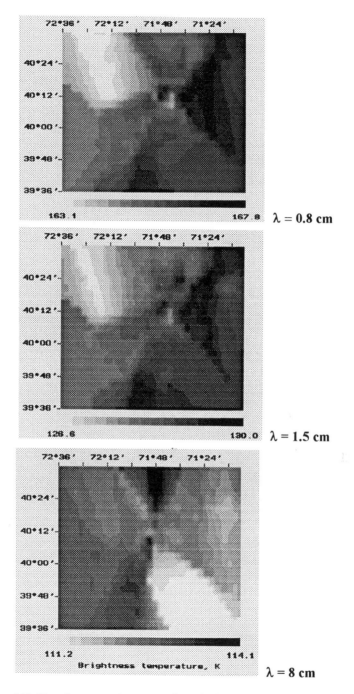

Figure 5.15. Three-frequency microwave radiometric signatures of the test area with oceanic internal waves. 2D-reconstruction was made during five flight legs (rosette) as shown in Figure 5.14 (JUSREX, 21 July 1992).

obviously non-stationary character, the question of spatial-time averaging of microwave signals becomes important to the analysis of the experimental data.

A qualitative interpretation of the microwave images can be speculated. According to the model of internal wave–current interaction, the roughness surface amplification processes may take place. Spatial non-uniformity of the surface current field may be causing the spatial modulations of the short gravity (decimeter and meter) surface waves' slopes on the one hand (effect of redistribution of the surface wave slopes), and/or deviations of the gravity–capillary wavenumber spectrum (effect of excitation of the high-frequency spectrum) on the other hand. Both effects are the cause of the variance in ocean microwave emission. Numerical estimations by using these models show that the values of brightness temperature contrast due to large-scale and small-scale roughness perturbations are about 3–5 K (at nadir viewing angle). A spatial distribution of the brightness contrast depends on a spatial distribution of the surface currents. In this context the effect of modulational secondary instability should also be considered. This effect may be causing the local short-time impluse-type perturbances of high-frequency wavenumber spectra. This burst-type mechanism gives local 'flashes' of radio-brightness that leads to the appearance of bright spots on microwave images. Although the hydro-dynamic aspect of the problem is not fully clear and requires additional numerical analysis, electromagnetic tasks can be solved successfully by using the combined methods of geometrical optics and small perturbations (see Chapter 3). It is important that internal wave–surface current interaction processes occur usually under the influence of wind, temperature and salinity. These environmental hydrophysical fields may mask the surface manifestation of oceanic internal waves observed by remote sensing sensors. In conclusion we should note that passive multifrequency microwave imagery will in future present a promising tool for aerospace operative diagnostics of internal oceanic processes and corresponding surface phenomena.

5.3 THE 'RELIC RAIN' SURFACE EFFECT

The rain leaves an original trace on the ocean surface, changing both the water salinity and the temperature, which also effects the formation of a thermohaline fine structure of the upper ocean layer [23, 51].

An interesting consequence of a short-period tropical shower is 'relic rain'. It is a temperature or salinity trace, registered through some time after the short-period tropical shower and due to the small-scale fluctuations in temperature [51] and salinity [52] in the surface layer of about 0.15 m thickness. The characteristic spatial and temporal scales of these fluctuations are 50–100 m and 10–15 min, respectively. The *in situ* measurements of 'relic rain', obtained by the authors [51] and [52], are shown in Figures 5.16 and 5.17.

Below we discuss the results of shipboard microwave observations of 'relic rain' phenomenon in the South China Sea [53]. The measurements were carried out in October 1984 during the 18th voyage of R/V *Professor Bogorov* by means of a multichannel microwave radiometer operating at frequencies of 20, 34, 37 and 48 GHz (vertical polarization) and 34 GHz (horizontal polarization). A microwave instrument was deployed on a gyroscopically stabilized turntable 10 m above the water line. The viewing angle was 75°. During the measurements a vessel was in drift. The wind speed was 4 m/s.

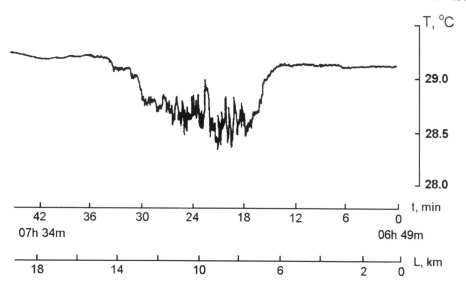

Figure 5.16. Sea surface temperature fluctuations due to 'relic rain' by A. Ginsburg *et al.* [51] on September 16, 1978, at 23°24′N, 80°43′W.

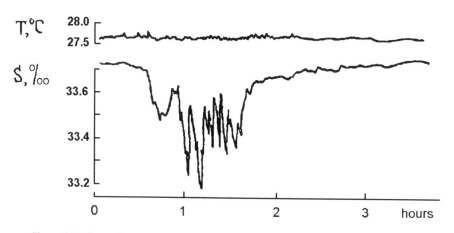

Figure 5.17. Sea surface salinity and temperature fluctuations due to 'relic rain' by M. Evans [52], May 1969, at 09°30′N, 119°00′W.

Figure 5.18 shows the synchronous recordings of sea surface microwave emission in the period from 16.30 to 24.00 of local solar time on October 17, 1984 near the geographic point of 12.5°N, 113.5°E. These recordings reflect both the process of short period tropical shower (*I*) and the 'relic rain' (*II*) effect on the sea surface. It should be noted that the separate cumulonimbus clouds appeared only at 17.00.

A shower of one hour's duration with an intensity of about 20 mm/h occurred at approximately 18.00, which led to a synchronous increase in brightness temperature T_b in all channels of the microwave radiometer at this time (Figure 5.18). The maximum

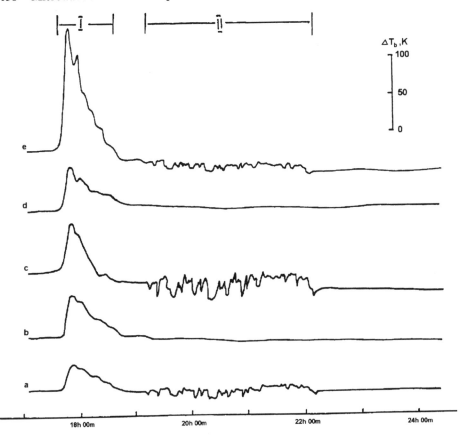

Figure 5.18. Brightness temperature variations due to both tropical shower (*I*) and 'relic rain' (*II*) on October 17, 1984, at 12.5°N, 113.5°E: (a) 48 GHz; (b) 37 GHz; (c) 34 GHz; (d) 20 GHz (all vertical polarization); and (e) 34 GHz (horizontal polarization).

contrast of T_b with horizontal polarization is $\Delta T_b = 150$ K, and with vertical polarization, $\Delta T_b = 20$–60 K. The recorded increments of T_b in this case are due entirely to atmospheric factors and related to the additional contribution of the microwave emission of precipitation reflected by the sea surface in the antenna direction. The substantial polarization difference in T_b is explained by a significant difference in the sea surface emissivity and reflectivity with different polarization at grazing viewing angles [54–56].

The spectral variations of sea surface microwave emission with a period of 5–15 min have been registered about through hour after the shower. These variations characterize the 'relic rain' phenomenon (see Figure 5.18). Unfortunately, the water temperature was measured only at a depth of 3 m by a probe located in the main machine water intake of the vessel, where the temperature fluctuations were not recorded. The water temperature was 29.8°C.

The variations in brightness temperature are observed at frequencies of 34 GHz (on both polarizations) and 48 GHz, and are virtually absent in the 20 and 37 GHz channels.

The changes in the contrast of sea surface brightness temperature in this case reach $\Delta T_b = 20$–30 K; there is both an increase and a decrease in T_b with respect to the background level.

So a large brightness temperature contrast in variations of sea surface microwave emission at a frequency of 34 GHz and the lack of any at a frequency of 37 GHz is quite surprising, and the question naturally arises, where is it real?

It should be noted that any changes in sea surface state at this time were not directly visible to the human eye similar to those as ripple water and white caps, observed during manifestations of oceanic internal waves (see section 5.2), although the sea surface brightness temperature variations exceed those caused by internal waves and reach 30 K.

It is clear that fluctuations in sea surface microwave emission due to 'relic rain' are not directly related to the changes in water temperature and salinity, because they are less than 0.7°C and 0.5‰, respectively, as well as meteorological variables of the atmosphere. Figure 5.19 shows the correlation between brightness temperature variations caused by 'relic rain' for two frequency channels, 34 and 37 GHz. For comparison, the calculated data according to (4.3) describing the brightness temperature changes, which would be due to the wind stress for wind speeds of 0, 5, 10 and 15 m/s, are also shown in Figure 5.19. The calculations were made for a viewing angle of 75°, where the increase in wind speed results in a decrease in sea surface brightness temperature [54–56]. The sea surface brightness temperature increment due to 'relic rain' at a frequency of 34 GHz exceeds that caused by wind stress in the 0–15 m/s range. A high correlation is demonstrated

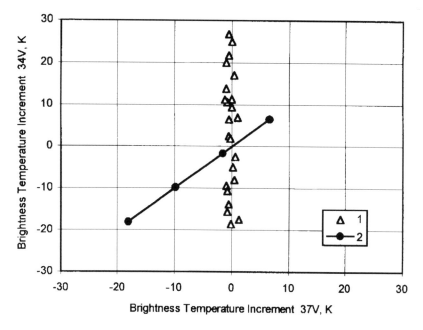

Figure 5.19. Regression dependence between sea surface brightness temperature variations at frequencies of 34 GHz and 37 GHz due to 'relic rain' (1—experimental data) and that which would be caused by the changes in wind speed from 0 to 15 m/s (—calculated data).

between brightness temperature variations for two channels combination of 34 and 37 GHz, if the wind stress is responsible for microwave emission changes. There is absolutely no correlation between brightness temperature variations due to 'relic rain', when the above-mentioned two frequencies are used, even though the difference between these two frequencies is very small.

Nowadays, the physical nature of the 'relic rain' phenomenon is unknown. As shown in [24, 57], the precipitation is the main reason providing the salinity fronts on the ocean surface, related to formation of fresh water isolated 'lenses'. During weak winds the rain of 1–2 hours' duration and 20 mm/h intensity and more can produce a 0.5–1.0‰ desalination of water layer thickness up to 1 m. A downturn in water temperature near the surface is often observed, which does not exceed 0.2–0.3°C [57]. The influence of changes in water temperature and salinity in the specified limits on water surface emissivity cannot explain high-contrast spectral variations in sea surface brightness temperature (see section 3.1).

Let us consider some processes of hydrodynamic mixing related to the changes in fine structure peculiarities of temperature and salinity fields in a subsurface ocean layer, which could have an effect on the spectrum of surface waves.

The sea water state equation in linear approximation is described in the following manner [58]:

$$\rho = \rho_0 (1 - \alpha \Delta T + \beta \Delta S) \tag{5.8}$$

$$\alpha = -\frac{1}{\rho}\left(\frac{\partial \rho}{\partial T}\right)_{S,P}, \qquad \beta = \frac{1}{\rho}\left(\frac{\partial \rho}{\partial S}\right)_{T,P}, \tag{5.9}$$

where ρ_0 is the *in situ* density of sea water, α the thermal expansion coefficient of sea water, β the saline contraction coefficient of sea water, and ΔT and ΔS the temperature and salinity deviations from *in situ* sea water values of T_0 and S_0, respectively.

According to (5.8), the temperature and salinity have an opposing effect on sea water density. Therefore, an increase in sea water density with depth is not a guarantee of hydrostatic stability of the water column. These are two types of instability due to double-diffusive convection [58]. If the temperature gradient produces the stabilization effect on the density distribution with the depth, but salinity is opposite, the convection instability of 'salt fingers' mode takes place. In the case where the salinity gradient produces the stabilization effect on the density distribution with the depth, but temperature is opposite, the convection instability of 'diffusive' mode is developed. Turner [59] described a series of laboratory experiments to study double-diffusive convection in both modes and compared this with published oceanic data to suggest explanations for some existing observations and to predict what might be measurable in future work.

Over the past few years there have been many observations of the fine structure and microstructure in the deep ocean which can be explained only in terms of 'double-diffusive' processes [60–64]. When there is a systematic association between temperature and salinity variations, with both properties increasing or decreasing together (so that they have opposing effects on the density), then it seems certain that the difference in molecular diffusivity of heat and salt in water can have a significant effect on the production of relatively well mixed layers and the transport of heat and salt between them. A monograph [65] describes many aspects of the applications of double-diffusive convection, including the oceanographic research.

After a short-period tropical shower the following situation occurs: the colder and fresher waters overlie the warm and salty waters. This promotes occurrence of hydrostatic instability and the development of double-diffusive convection in 'diffusive' mode.

We have carried out a laboratory experiment to study the double-diffusive convection in the tank [66] and found the relation between double-diffusion convection (both 'salt fingers' and 'diffusive') and the process of capillary wave generation on the water surface. According to [66], the double-diffusive convection at a depth of 10–15 cm results in transformation of the capillary wave spectrum for wavelengths less than 20 mm. Some spectral components of capillary waves changed their amplitude by more than one order. The degree of capillary wave spectrum transformation depends on the intensity of double-diffusion convection.

Taking into account the contribution of selective mechanisms into microwave emission of periodically rough water surface (3.3) and (3.4), it should be assumed that the high-contrast spectral variations in sea surface brightness temperature due to 'relic rain' could be explained by the transformation of wind capillary wave spectrum. According to (3.4) we can write the resonance conditions of microwave emission at wavelength λ by the periodical structure Λ of sea surface at viewing angle θ, as follows:

$$n\lambda = \Lambda(1 \pm \sin\theta) \quad n = 1, 2, \ldots . \tag{5.10}$$

Using equation (5.10) and experimental microwave data (Figure 5.18) we could evaluate the minimal relative band of rearrangement of capillary wave spectrum on the sea surface caused by 'relic rain'. Figure 5.18 shows that the two closest microwave spectral channels—in one the sea surface brightness temperature variations are detected and in the other they are not detected—are $f_1 = 34$ GHz and $f_2 = 37$ GHz respectively. Then, we obtain

$$\frac{\Delta\Lambda}{\Lambda} = \frac{(\lambda_1 - \lambda_2)}{\lambda_1}100\% = \frac{(f_2 - f_1)}{f_2}100\% \approx 7\%. \tag{5.11}$$

Now, we should estimate the resolution possibility of the microwave radiometer for detecting the capillary wave components on the sea surface. The value of $\Delta\Lambda$ depends on $\Delta\lambda \sim \Delta f_{\text{rad}}$, the radiometer bandwidth. Differentiating equation (5.10) we obtain

$$\Delta\Lambda = \frac{n}{1 \pm \sin\theta}\Delta\lambda.$$

Also, the value of $\Delta\Lambda$ depends on the antenna pattern beamwidth $\Delta\theta_1$, and the average slope of sea surface $\Delta\theta_2$ due to large-scale waves

$$\Delta\Lambda = \frac{n\lambda\cos\theta}{(1 \pm \sin\theta)^2}\Delta\theta_i \quad i = 1, 2.$$

Taking into account the real parameters of the microwave radiometer [67]: $f = 37$ GHz; ($\lambda = 8$ mm); $\Delta f_{\text{rad}} = 1$ GHz; $\Delta\theta_1 = 6°$ and setting $\Delta\theta_2 \leqslant 10°$ we finally obtain for a viewing angle of $\theta = 75°$.

$$\frac{\Delta \Lambda}{\Lambda} \leqslant 3\%,$$

while in observations at the nadir $\theta = 0$,

$$\frac{\Delta \Lambda}{\Lambda} \leqslant 20\%.$$

The instrumentation resolution error is small at grazing viewing angles, and increases with a reduction in viewing angle up to nadir. In our case the instrumentation resolution error has no influence on the estimated result of (5.11).

5.4 FRONTAL ZONE IN KUROSHIO REGION

The frontal zones are characterized by sharp spatial (vertical and horizontal) changes, first of all the temperatures, salinity, density and currents velocity in comparison with gradients of these characteristics in the environmental waters [24]. The variations in these parameters in the upper ocean layer change the electrodynamic properties of the surface. However, the ocean fronts, representing, generally speaking, the inclined layers, dividing waters with various thermohydrodynamic characteristics, are not always accompanied by the visible surface phenomena, for example, formation of the ripples or presence of temperature and salinity gradients on the ocean surface. Therefore, of great interest are the experimental results of microwave remote sensing of the Pacific sub-arctic front in Kuroshio region, where specific high-contrast variations in brightness temperature and backscattering cross-section of the ocean surface have been detected.

The microwave studies were carried out during the 16th voyage of R/V *Professor Bogorov* in September–October 1983 by means of a radiometer–scatterometer operating at a frequency of 37 GHz [38]. The microwave instrument was deployed on a gyroscopically stabilized turntable 10 m above the water line. The viewing angle was 75°. Remote sensing of the ocean surface was made on vertical polarization. The pattern of experiment included the meridional sections along 149 and 160°E in the region of the sub-arctic front, crossing Kuroshio and Kuril currents. The spatial areas of 10–60 km size, which are characterized by anomalous variations in ocean surface microwave parameters, were registered in the region of the front between 36 and 44°N. The lifetime of these ocean surface microwave anomalies was from some hours to 2–3 days.

Figure 5.20 shows the synchronous recordings of ocean surface brightness temperature T_b and backscattering cross-section σ in the period from 08.30 to 10.00 of local solar time on October 12, 1983, near the geographic point of 41°24′N, 149°00′E. The research vessel was moving at a velocity of 13 knots. The brightness temperature contrast reaches 30 K, and backscattering about 12 dB, the variations in T_b and σ are observed both in phase and antiphase, that is simultaneously increase in brightness temperature and back-scattering cross-section of the sea surface. The wind speed was 11 m/s and the weather was clear. It should be noted that any changes in sea surface state at this time were not directly visible to the human eye like ripple water and white caps, observed during manifestations of intensive oceanic internal waves (see section 5.2), although the brightness temperature variations reach 30 K, but backscattering cross-section about 12 dB. In

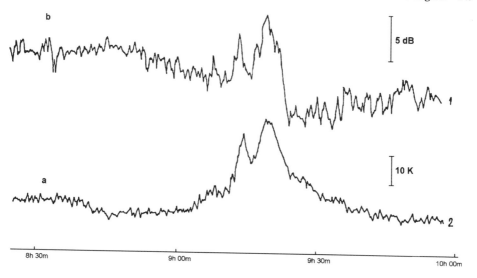

Figure 5.20. Shipboard record of ocean surface brightness temperature (a) and backscattering cross-section (b) in sub-arctic front of Pacific on October 12, 1983, near 41°24′N, 149°00′E. The ship's velocity is 13 knots.

some cases the horizontal changes in sea surface temperature with local gradient of 0.2–0.5°C/km according to *in situ* measurements were registered.

In Figure 5.21 are shown the synchronous recordings of the ocean surface brightness temperature T_b and backscattering cross-section σ in the period from 06.30 to 08.30 of local solar time on September 30, 1983 near the geographic point of 42°13′N, 160°00′E. In this case the area of 15 km size is characterized only by the increase in ocean surface brightness temperature about 24 K, while the level of backscattering cross-section remains practically constant. During measurements, the wind speed was 8 m/s, and the weather was cloudy—3–4 marks (stratocumulus).

Figure 5.22 shows the correlation between variations in brightness temperature and backscattering cross-section of the ocean surface caused by frontal zone processes. For comparison, the experimental data obtained by means of the same microwave instrument during observations of intensive oceanic internal waves (see section 5.2) are also shown in Figure 5.22. It can be seen that the influence of oceanic processes on microwave emission and backscattering of the ocean surface which take place in the frontal zone is quite different from those caused by oceanic internal waves. As described in section 5.2, the internal wave current modulates the ocean surface wave spectrum that gives rise to formation of rip on the surface and, in consequence, to the increase in backscattering cross-section and decrease in brightness temperature of the ocean surface. Nevertheless, the surface manifestation of processes in the frontal zone is characterized by highest brightness temperature contrast, as well as, backscattering cross-section of the ocean surface than those of internal waves and, in addition, they are accompanied by an increase in brightness temperature concurrently with the increase in backscattering cross-section of the ocean surface. It is assumed that the said variations in ocean surface

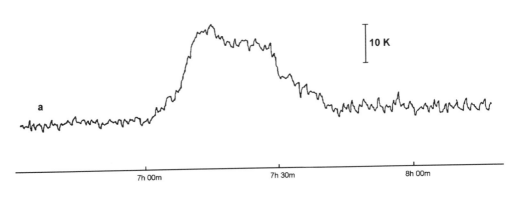

Figure 5.21. Same as Figure 5.20, but on September 30, 1983, near 42°13′N, 160°00′E.

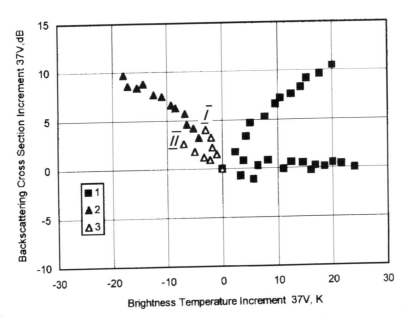

Figure 5.22. Regression between brightness temperature and backscattering cross-section variations in ocean surface at a frequency of 37 GHz due to: 1—frontal zone of Kuroshio; 2—internal oceanic waves; 3—wind gusts; the non-uniqueness (*I*) and (*II*) corresponds to $m = 1$ and $m = 2$ respectively for (5.17)–(5.19).

microwave characteristics in the frontal zone are not related to the influence of the current on the surface.

Let us consider the character of the influence of wind stress on microwave emission and backscattering of the ocean surface. Figure 5.23 shows the synchronous recordings of ocean surface brightness temperature T_b and backscattering cross-section σ obtained during investigations of wind gusts by means of the same microwave instrument [38]. These measurements were made during the 11th voyage of R/V *Professor Bogorov* in September, 1981, and were carried out to study the short-period variations of T_b and σ caused by wind gusts. The vessel was in drift. The average wind speed was 12 m/s, but the wind speed fluctuations ranged from 8 to 15 m/s with a period of from several seconds to 3–5 minutes. The wind gusts result in synchronous variations in brightness temperature and backscattering cross-section of the ocean surface with the same period. In this case the increase in backscattering cross-section of the ocean surface is accompanied by a decrease in brightness temperature (Figure 5.23). This fact agrees with theoretical and experimental data describing the wind stress influence on T_b and σ of the ocean surface under grazing viewing angles [54, 68, 69]. The correlation dependence between the increment of backscattering cross-section and brightness temperature of the ocean surface with wind gusts is shown in Figure 5.22. There is a non-uniqueness in correlation dependence between variations in T_b and σ. This effect could be explained by the non-stationary character of the wind-wave spectrum on account of fluctuations in the

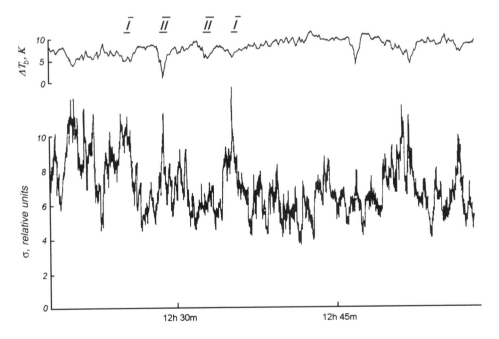

Figure 5.23. Shipboard record of ocean surface brightness temperature T_b and backscattering cross-section σ variations (vertical polarization) due to wind gusts on September 19, 1981, near 49°35′N, 160°20′E. The ship is in drift. (*I*) and (*II*) correspond to non-uniqueness in correlation dependence between variations in T_b and σ in Figure 5.22.

wind speed [70–73] and the high-order resonance phenomena in microwave back-scattering and emission of the ocean surface [74].

In general, we can see that correlation dependencies between T_b and σ, due to wind stress and internal oceanic waves, are qualitatively the same, while the correlation dependence between variations in T_b and σ, observed in frontal zone, is quite different.

To classify these microwave phenomena with respect to signal variations in the channels of the radiometer–scatterometer under grazing viewing angles, i.e. back-scattering cross-section $\sigma(t)$ and brightness temperature $T_b(t)$ of ocean surface, we will use the notion of interchannel derivative. According to a theorem on differentials [75], we can write the increment of backscattering cross-section changes in time as

$$\Delta\sigma = \frac{d\sigma(t)}{dt}\Delta t. \tag{5.12}$$

Similarly, we can write for the increment of brightness temperature changes in time

$$\Delta T_b = \frac{dT_b(t)}{dt}\Delta t. \tag{5.13}$$

Dividing (5.12) by (5.13) we obtain

$$\frac{\Delta\sigma}{\Delta T_b} = \frac{\left(\dfrac{d\sigma(t)}{dt}\right)}{\left(\dfrac{dT_b(t)}{dt}\right)}. \tag{5.14}$$

Expression (5.14) is a derivative of function, defined in the parametric form, and we can write

$$\Delta\sigma = \frac{d\sigma}{dT_b}\Delta T_b, \tag{5.15}$$

where $d\sigma/dT_b$ is purely an interchannel derivative.

We have the right to write the expressions (5.14) and (5.15) because the measurements of backscattering cross-section $\sigma(t)$ and brightness temperature $T_b(t)$ of the ocean surface by means of a radiometer–scatterometer have been combined in both time and space [38]. In a similar manner, the notion of interchannel derivative is also valid for measurements by means of a multichannel microwave radiometer described in section 4.1.

Let us analyze the regression diagram (Figure 5.22) by using expression (5.15). The variations in backscattering cross-section and brightness temperature of ocean surface under grazing viewing angles, caused by intensive oceanic internal waves, can be characterized as follows:

$$\left(\frac{d\sigma}{dT_b}\right)_{\text{Int.Waves}} < 0.$$

A similar result may be obtained for wind stress, although the interchannel derivative has non-uniqueness under the influence of wind gusts:

$$\left(\frac{d\sigma}{dT_b}\right)_{\text{Wind Stress}} < 0.$$

As for variations in backscattering cross-section and brightness temperature of the ocean surface in the frontal zone, we have a quite different result:

$$\left(\frac{d\sigma}{dT_b}\right)_{\text{Frontal Zone}} \geqslant 0.$$

To understand the reason for these very 'strange' variations in backscattering cross-section and brightness temperature of the ocean surface in the frontal zone, we first analyze the influence of wind gusts on the ocean surface. Figure 5.22 and Figure 5.23 show that the correlation dependence between synchronous variations in T_b and σ due to wind gusts has a non-uniqueness. It is known [76] that the radar signal is defined by the Bragg–Wolf resonance conditions of backscattering of electromagnetic waves λ by waves on the ocean surface Λ at given viewing angle θ as follows:

$$m\lambda = 2\Lambda \sin\theta, \quad m = 1, 2, \dots . \tag{5.16}$$

This means that in our case ($f = 37$ GHz, $\lambda = 8$ mm) the backscattering signal is determined at $\theta = 75°$ by the following surface components: $\Lambda \approx 4$ mm ($m = 1$); $\Lambda \approx 8$ mm ($m = 2$); etc. According to [54, 69], the change in intensity of the back-scattering signal is proportional to the change in the square of the amplitude of the surface wave Λ. Consequently, the change in magnitude of the signal scattered by the ocean surface may come from the change (depending on the disturbance conditions) in the amplitude of the surface wave Λ of the corresponding order m. In the given case the disturbance factor is a wind gust.

On the other hand, a change in the same spatial components leads to a change in the brightness temperature of the ocean surface (see Figure 5.23). According to (3.4) and (5.10), the microwave emission of the ocean surface also includes the 'resonance' nature, and we may combine equations (5.10) and (5.16) into the system

$$m\lambda = 2\Lambda \sin\theta,$$

$$n\lambda = \Lambda(1 \pm \sin\theta). \tag{5.17}$$

The system (5.17) has the solution $\theta \to 90°$

$$\Lambda = \frac{m\lambda}{2}, \quad m = n. \tag{5.18}$$

The correlation diagram obtained experimentally (see Figure 5.22) reflects the existence of the common solution (5.18) at least to the second-order inclusive ($m = 1; 2$) in the case of $\lambda = 8$ mm ($f = 37$ GHz), at $\theta = 75°$. Since the level of the backscattered intensity decreases with the increasing order, the direction of the scattering order m was determined from the decrease in $\Delta\sigma$. Consequently,

$$\left| \left(\frac{d\sigma}{dT_b} \right)^{I}_{\text{Wind Stress}} \right| > \left| \left(\frac{d\sigma}{dT_b} \right)^{II}_{\text{Wind Stress}} \right|. \tag{5.19}$$

However, from (5.19) it follows that the absolute value of the brightness temperature increment ΔT_b increases with increasing order n of the resonance phenomena.

So the 'responsibility' of the spatial component of Λ on the ocean surface for the resonance conditions of backscattering and emission of electromagnetic waves due to wind gusts follows from the observed quantization of the correlation relations of $\Delta \sigma$ and ΔT_b. On the other hand, it is just the wind gusts resulting in local non-stationarity of the capillary part of the wave spectrum on the ocean surface [70–73].

The analysis shows that transformation of the capillary wave spectrum on the ocean surface is responsible for the changing amount of the interchannel derivative, $d\sigma/dT_b$. If so, we can assume that the mesoscale variations of backscattering cross-section $\sigma(t)$ and brightness temperature $T_b(t)$ of the ocean surface, detected in the frontal zone, could also be related to the rearrangement of the capillary wave spectrum.

Below we discuss the hydrophysical processes that could be responsible for such peculiarity of variations in brightness temperature and backscattering cross-section of the ocean surface in the frontal zone.

The hydrological measurements taken during the 16th voyage of R/V *Professor Bogorov* were carried out by the Pacific Oceanographic Institute of Russian Academy of Sciences. Figure 5.24 presents a south–north meridional section in the temperature field

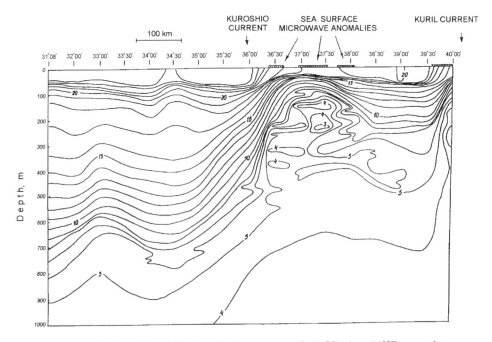

Figure 5.24. South–north meridional section in temperature field (°C) along 149°E across the Kuroshio current between October 9 and 12, 1983, by R/V *Professor Bogorov*.

along 149°E across the Kuroshio current between October 9 and 12, 1983. Hydrological stations of 1627–1643 were made through 30 miles from 31 up to 40°N. It should be noted that microwave remote sensing of the ocean surface was carried out round the clock in a continuous mode during the whole section.

In the upper part of Figure 5.24 the hatched rectangles on the X-axis indicate those locations where the above-mentioned variations in brightness temperature and back-scattering cross-section of the ocean surface were recorded. It can be seen that ocean surface microwave anomalies are located in the frontal zone of the Kuroshio current, where a distinct lens structure exists at a depth of 100–400 m (36–38°N), and also in the region of the Kuril current (39°30′–40°N). The horizontal scale of surface microwave anomalies is comparable to that of water lenses.

The vertical profiles of temperature, salinity and relative density (st. 1639) measured by CTD-sonde are shown in Figure 5.25. The *in situ* data point to a thermohaline structure of an intrusion origin. The vertical profiles are characterized by the presence of inversions of temperature and salinity. The thickness of the intrusion layers (both 'warm' and 'cold') ranges from 10 to 50 m. There are also inversions on the vertical profile of density which correspond to borders of the layers and characterize the hydrostatic instability there.

According to [23], the relation of vertical sizes of the fine structure peculiarities of temperature and salinity fields to horizontal ones equals 10^{-3}. Hence, the extent of the said intrusive layers must be within 10–50 km.

We can see that the anomalous changes in the microwave characteristics of the ocean surface in the frontal zone are registered just on the same spatial scales. That is, there is a certain kind of correlation between the horizontal sizes of intrusion lenses, being 100–400 m in depth and the development of hydrostatic instability there, and the spatial

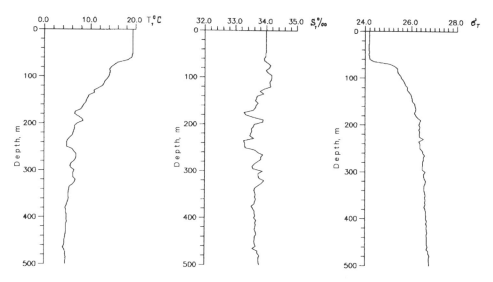

Figure 5.25. Vertical profile of temperature T, salinity S, and relative density σ_T at station No. 1639 of R/V *Professor Bogorov* on October 11, 1983, 38°00′N, 149°00′E.

areas on the ocean surface where the said variations of brightness temperature and back-scattering cross-section are detected.

It is difficult to give a complete physical explanation of this phenomenon. However, it is necessary to note that the intrusive stratification of water in the frontal zone creates favorable conditions for double-diffusion convection in its various forms and above/below intrusive interfaces [23]. The double-diffusion convection is just the secondary unstable process, which promotes the destruction of the initial intrusion, resulting in splitting them on numerous more thin intrusive and diffusive layers.

One of the best contemporaneous monographs [24] by the outstanding, now deceased, Russian oceanographer K. N. Fedorov is dedicated to the physical nature and fine structure of oceanic fronts. It summarizes our main physical knowledge about the thermohaline fine structure of the frontal origin:

1. Near-frontal zones are characterized by a predominance of intrusive character of thermohaline fine structure and most intrusive layers thickness (both warm and cold) range between 10 and 40 m, although there are separate intrusive layers of 1–5 m thickness, and in large-scale climatic frontal zones intrusive layers can attain a thickness of about 100–250 m.
2. The temperature inversions of cold and warm frontal origin intrusions are characterized by a vertical gradient of temperature ΔT from 0.02–0.04 to 1–3°C under quasipycnal compensations by appropriate gradient of salinity ΔS.
3. There are numerous indirect and for the present time a small field evidence of the presence of double-diffusion convection (both 'diffusive' and 'salt fingers' mode) at the upper and lower interface of intrusive layers and seams. Owing to convection of heat and mass on the borders, the lifetime of intrusions of 10–20 m thickness is usually limited by a period of some days.
4. The intrusive layers and seams cross the isopycnal surfaces in the direction of abundance or deficit of density (mass) relatively isopycnal balance. The presence of deficit or ambulance of density (mass) in the intrusions usually is well coordinated to direction of its evolution under influence of double-diffusion convection developing at its upper and lower interfaces.

The double-diffusive interleaving is discussed in the papers by McDougall [60, 61], where the linear stability analysis and finite amplitude, steady-state interleaving are presented. McDougall has reviewed numerous works dedicated to this problem and adjacent questions, and he gives a good list of references to the literature on it.

The background for developing of double-diffusive convection is characterized by the stability ratio R_ρ of the water column [23, 58, 60–62]. This ratio is based on the sea water state equation (5.8). If the water column is stabilized by the vertical distribution of temperature $(dT/dz < 0)$, and the unstable contribution is due to vertical distribution of salinity $(dS/dz < 0)$, R_ρ is determined by $R_\rho = \alpha \Delta T / \beta \Delta S$. For temperature inversion $(dT/dz > 0)$ and if the stabilization effect is caused by a rise in salinity with depth $(dS/dz > 0)$, R_ρ must be taken as $R_\rho = \beta \Delta S / \alpha \Delta T$.

If $R_\rho = 1$, the medium is characterized as indifferent stratification; $R_\rho > 1$ corresponds to hydrostatic stability; $R_\rho < 1$ corresponds to hydrostatic instability of the layer examined.

Figure 5.26 shows the distribution of R_ρ for the Northern Pacific and the regions where there are the favorable conditions for double-diffusive convection [77]. It can be seen that in the southerly from the sub-arctic front near to 41°N the double-diffusive convection takes place in the form of a 'salt finger', but in the northerly from the sub-arctic front the double-diffusive convection has developed in a 'diffusive' mode. Meanwhile, the area of the sub-arctic front includes both 'salt finger' and 'diffusive' modes of double-diffusive convection.

There are many areas of the Northern Pacific where the stability ratio $R_\rho < 3$, which characterizes the non-equilibrium oceanic medium as stable, but at the same time a medium may be on a threshold of stability or near it. This in turn provides the favorable conditions for working the amplification mechanisms (see section 5.1 and [23, 25]).

Thus, the analysis shows that the intensive thermohaline processes which take place in the frontal zone of the Kuroshio current could be responsible for the detected anomalous variations in brightness temperature and backscattering cross-section of the ocean surface. This ocean surface microwave effect is to some extent similar to that observed during 'relic rain', where its horizontal scale was 50–100 m and the depth of the thermohaline processes was limited by 15 cm (see section 5.3). As for the frontal zone, we observe the anomalous microwave phenomena on the ocean surface with a scale of 10–60 km, but the thermohaline processes are located at a depth of 100–400 m. In both cases the analysis shows that the high-contrast spectral variations in microwave emission due to 'relic rain' and anomalous variations in backscattering cross-section and brightness temperature of the ocean surface in the frontal zone could be related to the transformation of the capillary wave spectrum.

5.5 OCEANIC SYNOPTIC RING (ROSSBY SOLITON)

A central problem in oceanography is the study of synoptic eddies. The theoretical and experimental results concerning the generation and evolution of the oceanic synoptic eddies have been presented in a monograph [78]. The remote observation techniques of the oceanic eddies are different. The oceanic eddies are displayed on the visible and infrared images [79–83]. The non-homogeneity of the surface current fields changes the ocean surface radar cross-section. Therefore the ocean surface radar images allow for distinguishing the current boundaries [84]. The altimetry method allows us to detect sea level variations due to synoptic eddies and Rossby waves [85–88]. Below, we discuss the applications of the passive microwave technique for observations of the oceanic synoptic eddies, including deep water ones, with the ocean surface emission [89, 90].

The joint experiment in studying the Pacific sub-arctic frontal zone along the 149°E meridian from 34 to 45°N was carried out by the research vessel *Academician Lavrentev* and a TU-134 research aircraft during October 1990. The experiment involved both microwave remote sensing of the ocean surface and *in situ* measurements. The pattern of the experiment is shown in Figure 5.27. A unique oceanic phenomenon was detected during this mission. It is a synoptic Kuroshio Ring, consisting of two coupled oceanic eddies of opposite sign—cyclone and anticyclone.

Airborne microwave mapping of the ocean surface was carried out by means of a multichannel microwave imaging radiometer [67], on October 18, 20 and 21. The aircraft

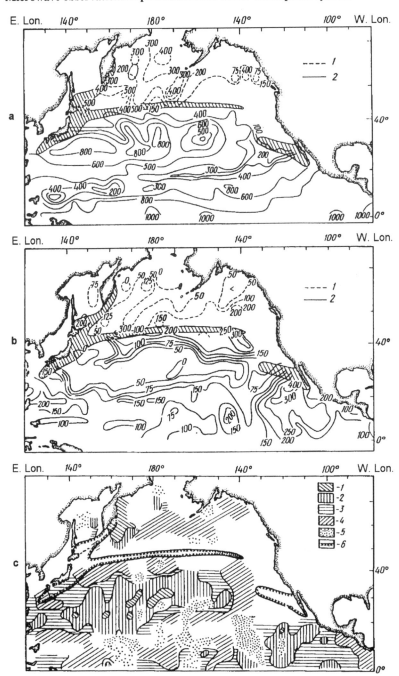

Figure 5.26. Topography of double-diffusive convection in the Northern Pacific [77]. Under boundary (a) and upper boundary (b) of the layers corresponding to 'diffusive' (*1*) and 'salt fingers' (*2*) mode, and distribution of R_ρ (c): *1*—$1 < R_\rho \leqslant 2$; *2*—$2 < R_\rho \leqslant 3$; *3*—$3 < R_\rho \leqslant 5$; *4*—$R_\rho > 5$; *5*—no data; *6*—mixed type of processes.

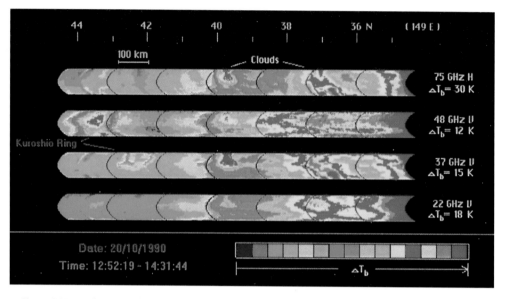

Figure 5.28. Multispectral microwave images of ocean surface along 149°E meridian on October 20, 1990, by TU-134 research aircraft at frequencies of 22, 37, and 48 GHz (vertical polarization) and 75 GHz (horizontal polarization). The Kuroshio Ring is seen at frequencies of 37 GHz and 48 GHz only.

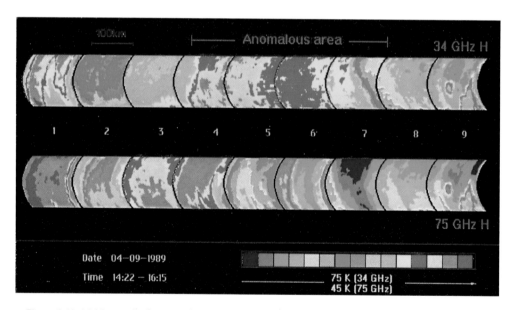

Figure 5.41. Multispectral microwave images (section) of Sea of Okhotsk on September 4, 1989, by IL-18 research aircraft at frequencies of 34 and 75 GHz (horizontal polarization). The position of frames (1, 2,...,9) along the aircraft track is shown in Figure 5.40.

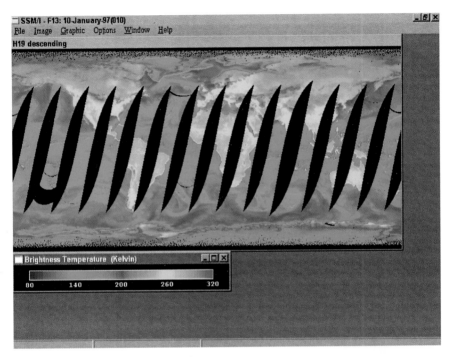

Figure 5.44. Global brightness temperature from SSM/I DMSP F13 at a frequency of 19 GHz, horizontal polarization, on January 10, 1997. Descending passes, equator crossing: 05:45 local solar time.

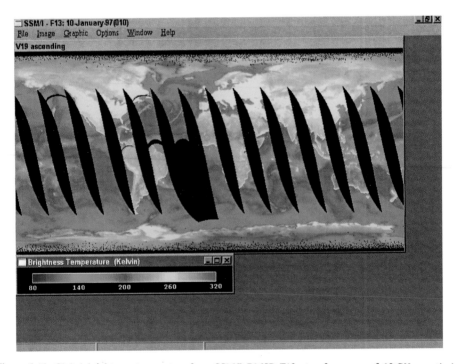

Figure 5.45. Global brightness temperature from SSM/I DMSP F13 at a frequency of 19 GHz, vertical polarization, on January 10, 1997. Ascending passes, equator crossing: 17:45 local solar time.

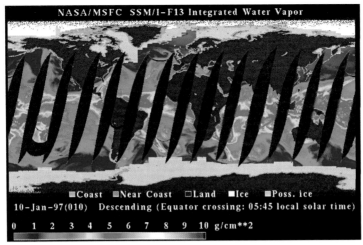

Figure 5.46. Integrated water vapor over the ocean, on January 10, 1997.

Figure 5.47. Cloud liquid water content over the ocean, on January 10, 1997.

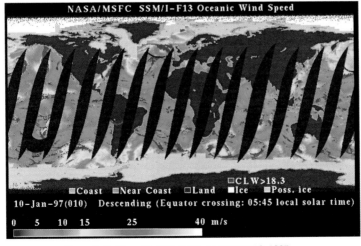

Figure 5.48. Oceanic wind speed, on January 10, 1997.

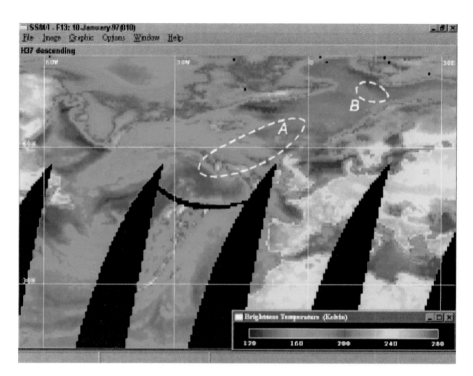

Figure 5.49(a). Microwave image of Northern Atlantic from SSM/I DMSP F13, on January 10, 1997, at a frequency of 37 GHz, horizontal polarization. Descending passes, equator crossing: 05:45 local solar time.

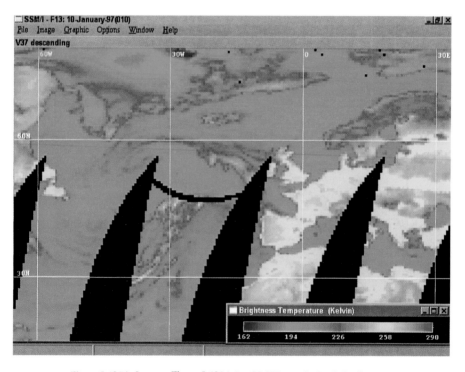

Figure 5.49(b). Same as Figure 5.49(a), but 37 GHz, vertical polarization.

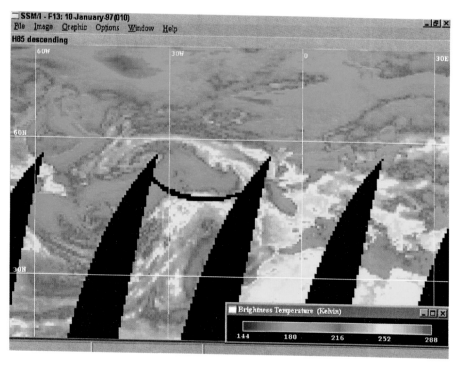

Figure 5.49(c). Same as Figure 5.49(a), but 85 GHz, horizontal polarization.

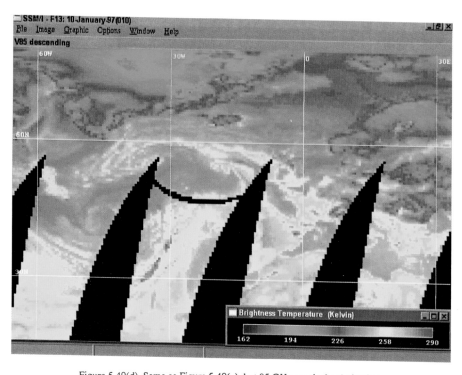

Figure 5.49(d). Same as Figure 5.49(a), but 85 GHz, vertical polarization.

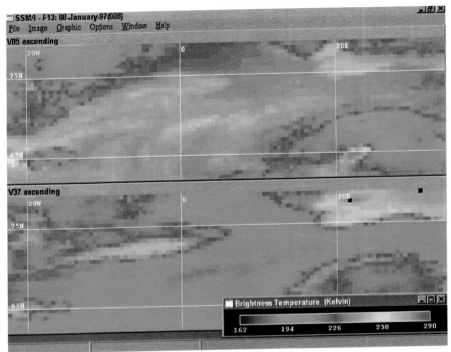

Figure 5.52(a). Microwave images of Norwegian Sea from SSM/I DMSP F13 at frequencies of 37 GHz (down) and 85 GHz (upper), both vertical polarization, on January 8, 1997, ascending passes, equator crossing: 17:45 local solar time.

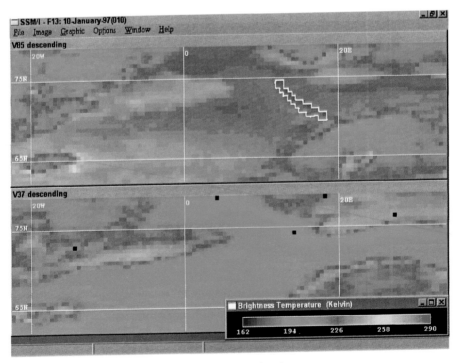

Figure 5.52(b). Same as Figure 5.52(a), but on January 10, 1997, descending passes, equator crossing: 05:45 local solar time. White bordered area corresponds to $\left(dT_b^{85V} \middle/ dT_b^{85H} \right) < 0$.

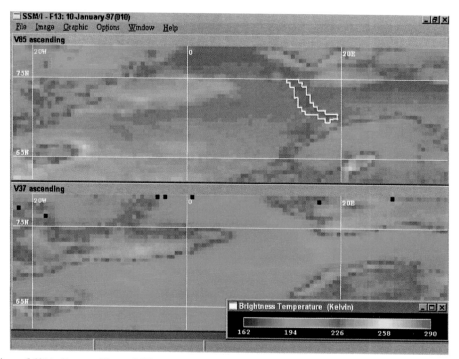

Figure 5.52(c). Same as Figure 5.52(a), but on January 10, 1997, ascending passes, equator crossing: 17:45 local solar time. White bordered area corresponds to $\left(dT_b^{85V} \Big/ dT_b^{85H} \right) < 0$.

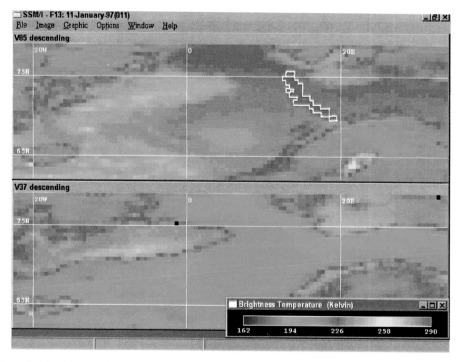

Figure 5.52(d). Same as Figure 5.52(a), but on January 11, 1997, descending passes, equator crossing: 05:45 local solar time. White bordered area corresponds to $\left(dT_b^{85V} \Big/ dT_b^{85H} \right) < 0$.

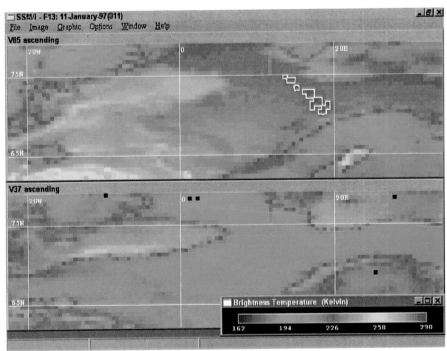

Figure 5.52(e). Same as Figure 5.52(a), but on January 11, 1997, descending passes, equator crossing: 17:45 local solar time. White bordered area corresponds to $\left(dT_b^{85V} \middle/ dT_b^{85H} \right) < 0$.

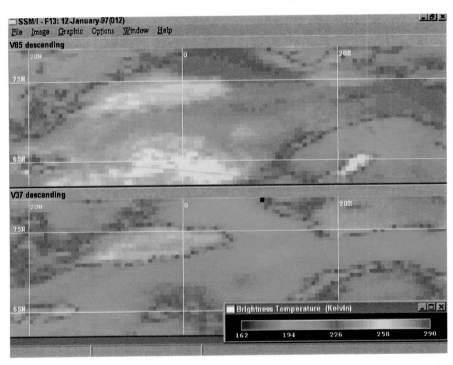

Figure 5.52(f). Same as Figure 5.52(a), but on January 12, 1997, descending passes, equator crossing: 05:45 local solar time.

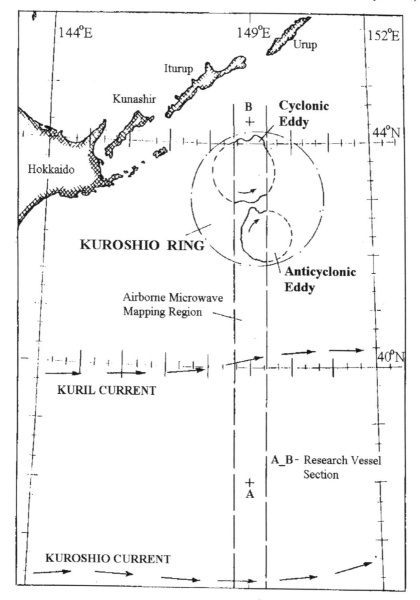

Figure 5.27. Pattern of experiment on joint study of the Kuroshio Ring by TU-134 aircraft and the research vessel *Academician Lavrentev* during October, 1990. The size and position of the anticyclonic and cyclonic eddies are shown according to airborne microwave data.

flight height was 10 km and the swath width was 60 km. Multispectral microwave images of this region at frequencies of 22, 37 and 48 GHz (vertical polarization) and 75 GHz (horizontal polarization), obtained on October 20, are shown in Figure 5.28 (in the colour section).

At the same time, the research vessel made a hydrological section along 149°E meridian, taking hydrological observations every 20 miles from 38°00′ to 44°20′N, and a continuous echo-radar recording of the acoustic non-homogeneities in the upper ocean layer was carried out [91, 92]. In addition, standard meteorological observations were carried out on board, and the water surface temperature was measured. The research vessel worked on the section during a period from October 16 to 18. The depth distributions of relative density, temperature and salinity of the upper ocean layer (0–300 m) are shown in Figure 5.29. The hydrologic measurements present a pattern, which is typical for the region under investigation. The region inspected covers the northern part of the Pacific Ocean sub-arctic zone, traversing the northern section of the sub-arctic front at a latitude of approximately 40°30′N, and ends in the interfrontal zone, not reaching a southern section.

As can be seen in Figure 5.29, the section traversed two synoptic anticyclonic eddies which are a common phenomenon for this region. These eddies are formed from the Kuroshio current meanders and run north along the continental slope, remaining preserved sometimes over 1.5–2 years [93]. In the sections presented on density, temperature and salinity fields, the anticyclonic eddies are characterized by distortions of the isolines in the form of a funnel. One of the eddies occupies the area of 42–44°N, and another eddy occupies an area to the south of 39°20′N. In the case discussed we are more interested in the anticyclonic eddy located in the northern part of the section, which is coupled with a cyclonic eddy. The cyclonic eddy is characterized by the distortions of the

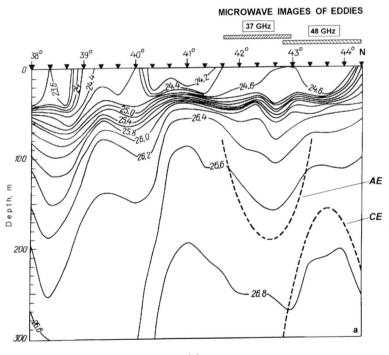

(a)

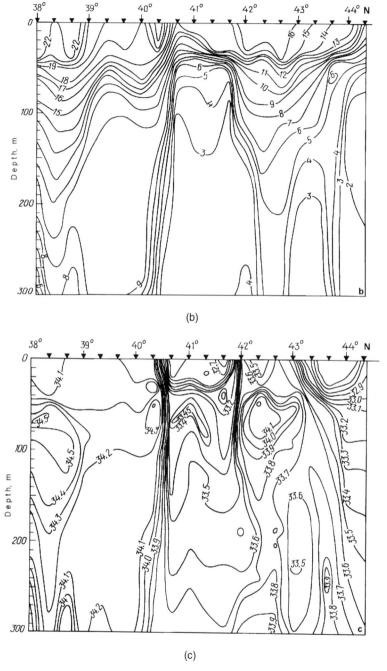

(b)

(c)

Figure 5.29. South–north meridional section along 149°E across the Kuroshio Ring between October 16 and 18, 1990, by R/V *Academician Lavrentev*: (a) relative density, dashed line indicates the anticyclonic eddy (AE) and cyclonic eddy (CE); (b) temperature (°C); (c) salinity (‰).

isolines in the form of a cupola (see Figure 5.29). The anticyclonic eddy is disposed near the surface, but the cyclonic eddy is disposed at a depth of more than 150 m.

We consider the microwave images at some frequencies to demonstrate the spectral peculiarities of sea surface brightness temperature due to the Kuroshio Ring (Figure 5.28, colour section). The area of the positive brightness temperature contrast of 6–7 K and size about 200 km is clearly seen on the image near 43–44°N at frequency 48 GHz only. At the same time the area of 150 km is seen only at frequency 37 GHz near 42–43°N and has a brightness temperature contrast of about 3–4 K. Both areas are located at the position of the Kuroshio Ring (see Figure 5.29(a)).

The microwave image of the Kuroshio Ring at 37 GHz coincides with the position of the surface warm core of the anticyclonic eddy (see Figure 5.29(b)). But the microwave image of the Kuroshio Ring at 48 GHz coincides with the size and position of the cyclonic eddy, which is disposed at a depth of more than 150 m (see Figures 5.29(a), (b), (c)).

It is significant that both eddies are clearly seen in microwave images and in addition the brightness temperature contrast due to the deep water eddy is much greater than one, caused by the near-surface eddy. Two coupled oceanic eddies of opposite sign—cyclone and anticyclone—are called a Rossby soliton. This problem has been studied in much detail by means of numerical simulation [94–96]. The experimental data of remote sensing (Figure 5.27 and Figure 5.28, colour section) and *in situ* measurements (Figure 5.29) are in good agreement with numerical modeling of the Rossby soliton by Kizner [96] (Figure 5.30). The possibility of using the microwave technique for detecting simultaneously both eddies is very important for oceanography, because the position of one eddy with respect to another defines the general direction of their propagation in the ocean [78].

The weather conditions during microwave observations of the Kuroshio Ring were: near surface wind speed—7 m/s, weather state—clear. South of 41°N was cloudy. The cumulonimbus clouds are also seen on the microwave images (see Figure 5.28, colour section), but clouds increase the brightness temperature in all channels in a way that is quite different from that caused by the Kuroshio Ring.

The regression diagrams are used (see Figure 5.31) to analyze the brightness temperature variations in the region of the Kuroshio Ring shown in Figure 5.28 (colour section). Brightness temperature variations are due to both oceanic processes and meteorological variables of the atmosphere. Calculated brightness temperature data (three points), based on *in situ* shipboard measurements (see Table 5.3), are also presented in Figure 5.31. The calculations were made for a viewing angle of 75° and aircraft flight height of 10 km, using (4.3).

A high correlation is demonstrated between brightness temperature variations for a combination of two channels, 22 and 75 GHz (Figure 5.31). In this case, the meteorological variables—humidity, cloud liquid water content, air and water surface temperature—are mainly responsible for brightness temperature changes. The experimental and theoretical data are in good agreement. The similar highly correlated relationship between brightness temperature variations also takes place in the diagrams for a combination of both frequencies, 22 and 37 GHz, and 22 and 48 GHz.

At the same time, there is absolutely no correlation between brightness temperature

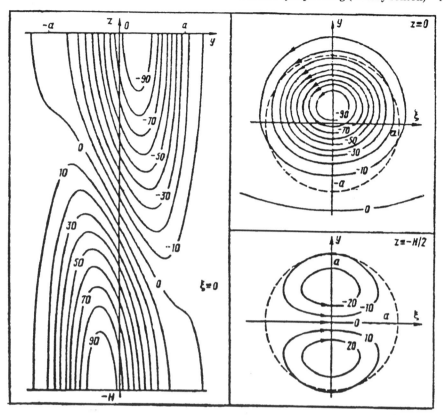

Figure 5.30. Numerical result of Rossby soliton by Z. Kizner [96]. Stream function isolines (in %
from maximum) in vertical and horizontal planes.

Table 5.3. Meteorological parameters used for brightness temperature
calculation

Parameter	Point 1	Point 2	Point 3
Sea surface temperature (°C)	16	19	19
Wind speed (m/s)	10	11	11
Temperature of air (°C)	14.5	18	18
Absolute humidity (g/m³)	9.6	12.3	12.3
Cloud liquid water (kg/m²)	—	—	0.4

variations due to oceanic eddies. This situation corresponds to the cyclonic eddy, when a
frequency of 48 GHz is used, and to the anticyclonic eddy, when a frequency of 37 GHz
is used. The brightness temperature contrast at a frequency of 48 GHz due to cyclonic
eddy reaches 6–7 K, and to be comparable with one caused by the clouds system at that

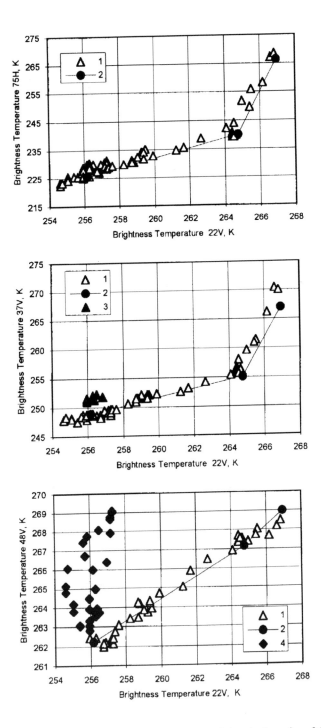

Figure 5.31. Regression between brightness temperature variations in the region of the Kuroshio Ring at frequencies of 22 GHz and 37, 48, 75 GHz according to Figure 5.28 (colour section): 1, 2—brightness temperature variations due to meteorological parameters (1—experimental data, 2—calculated data); 3—brightness temperature variations due to anticyclonic eddy; 4—brightness temperature variations due to cyclonic eddy.

frequency. On the other hand, as follows from regression diagrams, this spectral effect in upwelling microwave emission of the ocean–atmosphere system cannot be related to the atmosphere meteorological variables, as well as, sea surface temperature and wind stress. It is purely the result of the influence of the deep oceanic processes on the surface, in particular caused by synoptic eddies.

Based on the notion of the interchannel derivative (5.14) and (5.15), we can write for a combination of two frequencies, 22 and 48 GHz (Figure 5.31), in the case of brightness temperature variations due to meteorological variables

$$\left(\frac{dT_b^{48V}}{dT_b^{22V}} \right)_{\text{Meteo Var}} \approx 0.5,$$

and for brightness temperature variations due to cyclonic eddy

$$\left(\frac{dT_b^{48V}}{dT_b^{22V}} \right)_{\text{Cyclonic Eddy}} \to \infty.$$

It should be noted that the interchannel derivative for other combinations of frequencies, 22 and 37 GHz, and 22 and 75 GHz in the case of brightness temperature variations due to meteorological variables is also positive and has a finite value

$$\left(\frac{dT_b^{37V}}{dT_b^{22V}} \right)_{\text{Meteo Var}} \approx 0.8\text{--}2, \quad \text{and} \quad \left(\frac{dT_b^{75H}}{dT_b^{22V}} \right)_{\text{Meteo Var}} \approx 1.9\text{--}10.$$

The ocean surface microwave phenomenon caused by the Kuroshio Ring is similar to those described in sections 5.3 and 5.4, but there is a difference in the spatial scale. The size of spectral brightness temperature variations on the ocean surface due to 'relic rain' was 50–100 m, and frontal zone, 10–60 km. In our case, we deal with the anomalous microwave effect on the ocean surface, reaching 150–200 km.

Let us consider in more detail how the microwave images of the eddies correlate with internal structure and with the physical temperature and salinity distribution. From a comparison of Figures 5.28 (colour section) and 5.29 it can be seen that the area of positive brightness temperature contrast at 37 GHz, in its disposition and spatial size, coincides with a warm core of the anticyclonic eddy whose center is approximately at 42°30′N, where a funnel-shaped distribution of the isolines is observed at a depth to 300 m. The other, substantially greater brightness temperature contrast at 48 GHz relates to the north periphery of the anticyclonic eddy. Here, elevated values of the temperature, salinity and density horizontal gradients are observed in the upper layer (see Figure 5.29), whereas at depths of 150–300 m a dome-shaped distribution of the isolines is observed, which suggests the existence of a zone with cyclonic circulation of water masses.

The gradient of physical temperature on the surface within the confines of an eddy is 3–5°C at a distance of 300 km; consequently, a brightness temperature contrast of the anomaly in the 37 GHz channel may be caused by changes in the surface physical temperature. The absence of a similar anomaly in a neighboring channel might be explained by an appreciable attenuation of radiation in the atmosphere at the above-stated frequencies. However, the character of an anomaly manifestation on the surface and its spatial configuration indicate that it is most likely caused here by different factors. Such a conclusion is all the more valid with respect to the spatial structure in the 48 GHz channel, where the recorded levels of changes in the brightness temperature greatly exceed the physical temperature variations of a surface.

The internal structure of the eddies under consideration manifests itself more clearly in the section of temperature and salinity field (see Figure 5.29(b), (c)). Here, the temperature and saline cores of the eddies are easily distinguished. These cores are defined from both sides by clearcut local fronts at 42 and 44°N, with high values of the horizontal gradient. The microwave anomalies on the ocean surface are just located within these fronts (see Figure 5.28, colour section). But the salinity gradients, as well as the temperature ones, cannot be the direct cause of the brightness temperature contrast observed here. It is most likely that such an effect is produced by a shear current, which inevitably presents in these fronts (Figure 5.32). The shear current can have an effect on the surface, either directly or indirectly, through instabilities developing in a region of increased value of the velocity gradient.

Ocean surface microwave anomalies also prove to be related to dynamic disturbances in the near-surface layer. These anomalies coincide with an area of the eddies' interaction

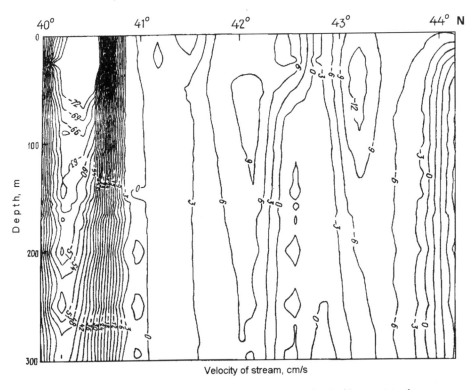

Figure 5.32. Velocity distribution of geostrophic currents calculated with respect to the zero surface on the 1000 m horizon [90]. Negative velocity values denote the eastward stream; positive values denote the westward stream.

with the jet flow of colder and desalinated sub-arctic water, which is adequately traced with isohalines (see Figure 5.29(c)) and can be seen in the velocity field of geostrophic flows calculated with respect to zero surface on the 1000 m horizon (see Figure 5.32). It should be pointed out that the actual value of the orbital velocity of flow in such eddies, estimated from the results of instrumental measurements, is of the order of 1 m/s, but its progressive motion velocities are of the order of 0.01 m/s [97].

The character of micro-scale processes within the scope of the anomalous area under consideration can be judged from Figure 5.33. It shows an echo-radar record of acoustic non-homogeneities in the upper layer to 150 m, obtained as a ship moved along the section. The lower part of the image in this figure is a continuation of the upper one, and on the whole the image covers a portion of the section at 43°30′–44°10′N and comprises the north boundary of a surface anomaly observed in the 48 GHz channel. Thin-layer non-homogeneities grouped in the form of characteristic frontal zones with developed micro-scale structure of hydrophysical fields are of interest here. Usually, they are related to processes of intrusive interaction at the boundaries of heterogeneous water masses and point to the presence of advection which maintain the horizontal gradients of temperature and salinity field in an enhanced state.

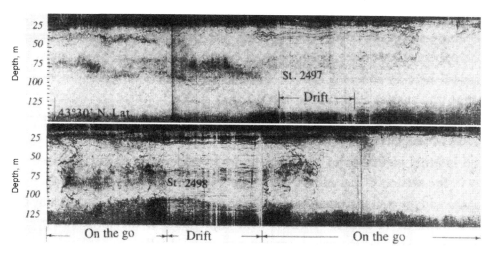

Figure 5.33. Echo-radar record of non-homogeneities in the water medium along 149°E meridian, obtained by remote acoustic probing of research vessel *Academician Lavrentev* between 43°30′N and 44°10′N, on October 18, 1990. The lower part is a continuation of the upper one.

5.6 SURFACE EFFECT FROM ORIGIN OF TROPICAL CYCLONE *WARREN*

The influence of typhoons and hurricanes on the ocean upper layer during their motion over a water surface has been studied in great detail [98–105]. At the same time, the detailed investigations of the interaction between ocean and atmosphere on the eve of a tropical cyclone's (TC) origin are still rare, as its planning is complicated by the unreliable prediction of exact places of tropical cyclone formation. For this reason any data about the disturbance of the active ocean layer, resulting in the final occurrence of a tropical cyclone, till now have been the subject of great interest.

During the 18th voyage of R/V *Professor Bogorov*, the microwave remote sensing of the ocean surface during the origin of tropical cyclone *Warren* in the South China Sea was made [106]. The formation of the cyclone was observed on October 20–31, 1984. The stages of development of tropical cyclone *Warren* are shown in Table 5.4 according to the data of the Japanese Meteorological Center. The shipboard microwave measurements were carried out during October 16–25 near the geographic point of 12.5°N and 113.5°N at a distance of 250–400 km from the cyclone's center. The microwave remote sensing of the ocean surface was made by means of a multichannel radiometer at frequencies of 20, 34, 37 and 48 GHz with vertical polarization [67]. A microwave instrument was deployed on a gyroscopically stabilized turntable 10 m above the water line. The viewing angle was 75°. The measurements were conducted in continuous mode over 10 days, while the vessel was mainly in drift.

Table 5.4. Position, wind speed, pressure, and classification of tropical cyclone *Warren* (8423)

Date	Position	Wind speed, m/s	Pressure at center, mb	TC classification
17/10/84	12.5°N, 113.5°E			High-contrast spectral variations in ocean
18/10/84	Geographic point of microwave observations	4–6	—	surface brightness temperature at
19/10/84	of ocean surface			microwave frequencies
20/10/84	12.0°N, 118.0°E	10	1006	Tropical disturbance
21/10/84	12.5°N, 117.8°E	10	1006	Tropical disturbance
22/10/84	12.2°N, 117.5°E	12	1004	Tropical disturbance
23/10/84	13.5°N, 116.5°E	15	998	Tropical disturbance
24/10/84	14.3°N, 115.5°E	23	990	Tropical storm
25/10/84	14.5°N, 114.7°E	27	980	Strong tropical storm
26/10/84	14.9°N, 114.6°E	30	975	Strong tropical storm

The anomalous variations in ocean surface microwave emission were detected during the three days prior to the origin of the tropical cyclone. The tropical disturbance, which is the earliest stage of the tropical cyclone, was classified on October 20. But the high-contrast spectral variations in ocean surface brightness temperature T_b with a period of 5–8 hours were observed on October 17, 18 and 19.

Figure 5.34 shows the synchronous recordings of ocean surface microwave emission in the period from 08.30 to 16.30 of local solar time on October 17, 1984. It is interesting to note that the brightness temperature of the ocean surface at a frequency of 34 GHz increases by 20 K, but the brightness temperature at a frequency of 48 GHz decreases by 30 K. At the same time the level of brightness temperature at frequencies of 20 and

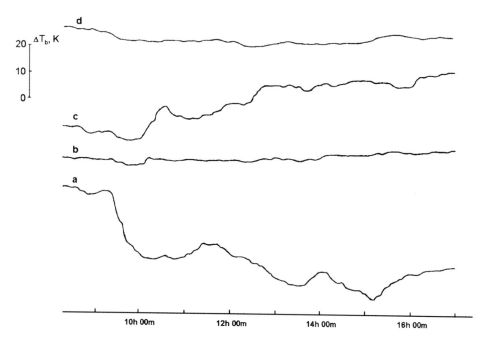

Figure 5.34. Brightness temperature variations of ocean surface on the eve of TC *Warren*'s origin on October 17, 1984, at 12.5°N, 113.5°E: (a) 48 GHz; (b) 37 GHz; (c) 34 GHz; (d) 20 GHz (all vertical polarization).

37 GHz remains practically invariable. The said variations in brightness temperature were detected during daylight hours. The sea surface temperature was 29.8°C, and the wind speed was 4–5 m/s; the weather condition was clear.

Figure 5.35 shows the correlation between brightness temperature variations for two frequency channels, 34 and 48 GHz. For comparison, the calculated data according to (4.3) for the said viewing angle and describing the brightness temperature changes which would be due to wind stress for wind speeds of 0, 5, 10 and 15 m/s are also shown in Figure 5.35. Using the concept of interchannel derivative ((5.14) and (5.15)) we can see that for the said two channels the changes in ocean surface brightness temperature on the eve of the tropical cyclone origin are described by the negative interchannel derivative

$$\left(\frac{dT_b^{34V}}{dT_b^{48V}}\right)_{TC} < 0,$$

while the change in brightness temperature, which could be due to wind stress, has positive interchannel derivative

$$\left(\frac{dT_b^{34V}}{dT_b^{48V}}\right)_{Wind\ Stress} > 0.$$

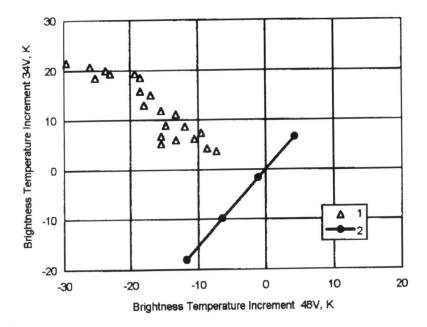

Figure 5.35. Regression dependence between ocean surface variations of T_b at frequencies of 34 GHz and 48 GHz on the eve of TC *Warren*'s origin (1—experimental data), corresponding to Figure 5.34 and variations in T_b, which would be due to the changes in wind speed from 0 to 15 m/s (2—calculated data).

The said peculiarities of spectral variations in ocean surface brightness temperature at frequencies of 34 and 48 GHz exclude their relation to wind stress and meteorological variables because the microwave measurements were conducted at constant conditions of sea state, wind speed, and clear weather.

We suppose that this microwave phenomenon could be related to the oceanic processes taking place on the eve of the tropical cyclone's origin. Actually, it is known [98, 101, 105, 107] that tropical cyclones originate only over the ocean and reach hurricane or typhoon status if the water temperature is more than 26°C through the depth up to 60 m. In the large-scale interaction of ocean and atmosphere the important relation link, determining a thermal state of atmosphere, is the redistribution of thermal energy in the ocean due to general water circulation. One of the number of processes which, by virtue of its spatial spreading and practically unlimited time of action in the ocean, can play a role in the climate-forming factor is double-diffusive convection in 'salt fingers' mode [108]. The relation of this process with the formation of the large-scale structure of waters in the ocean and with climatic conditions is shown in the literature [65, 77, 109–111].

In the Northern Pacific the regions where tropical cyclones originate [102] coincide with zones where there are favorable conditions for salt-finger convection at a depth of up to 400–800 m [77, 108] (Figure 5.36).

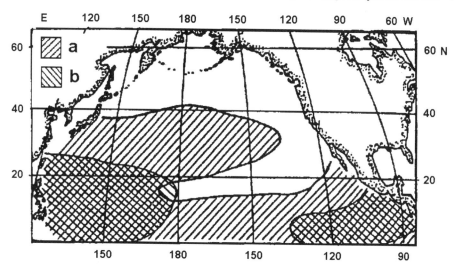

Figure 5.36. Zones (a) where there are favorable conditions for salt-finger convection at a depth of up to 400–800 m [77, 108] and regions (b) where tropical cyclones originate [102].

It is known that the flow of heat and moisture from the ocean to the atmosphere is a source of energy for tropical cyclones. The flow intensity depends on the heat reserve of the active ocean layer, which is determined by a thermal potential [101]. In the region of tropical cyclone origination, where the wind speed is still low, the thermohaline processes of the upper ocean layer play a significant role in vertical heat and mass transfer [23]. It is significant that double-diffusive convection in salt-fingers mode provides the energy transfer against the density gradient [23, 65].

The detected high-contrast spectral variations in ocean surface brightness temperature over a period of 3 days could characterize the synoptic process of energy exchange between ocean and atmosphere, resulting in variability of the active ocean layer. The manifestation of that process in the microwave emission of ocean surface to some extent is similar to those observed in the Kuroshio region (see sections 5.4 and 5.5). It should be noted that purely a 'relic rain' phenomenon has been detected on the eve of origin of tropical cyclone *Warren*. The recordings of the ocean surface microwave emission shown in Figure 5.18 are a continuation of those in Figure 5.34.

Thus, we can see that the many oceanic processes from the small scale up to synoptic ones are characterized by the spectral variations in ocean surface brightness temperature, which in turn point to a transformation of capillary wave spectrum (see sections 5.3 and 5.4).

Now, using the resonant theory of microwave emission of the small-scale rough water surface, described in section 3.2, and computer simulation input we try to explain the spectral variations in ocean surface brightness temperature shown in Figure 5.34. In the one-dimensional case the brightness temperature contrast, caused by a disturbance of the spectrum of small-scale roughness $F(k)$, can be calculated by using formula (3.5):

$$\Delta T_{\rm b} \approx T_0 k_0^2 \int G\left(\frac{k}{k_0}\right) F(k)\, dk. \tag{5.20}$$

We assume that the undisturbed spectrum of small-scale roughness $F(k)$ on the ocean surface is

$$F(k) = Ak^{-p}, \tag{5.21}$$

where $A = 10^{-2}$, $p = 2.8$, which corresponds to a wind speed of m/s.

To obtain the brightness temperature contrast at four frequencies of 20, 34, 37 and 48 GHz (all vertical polarization) as shown in Figure 5.34, the computer simulation results in the following modified spectrum of small-scale roughness (see Figure 5.37)

$$F_{\rm M}(k) = Ak^{-p}\left(1 + \sum_{i=1}^{m} b_i \exp\left(-c_i(k - k_{si})^2\right)\right), \tag{5.22}$$

where $m = 4$, $b_1 = 1.16 \cdot 10^3$, $b_2 = 200$, $b_3 = 150$, $b_4 = 40$, $c_1 = c_2 = c_3 = c_4 = 20$, $k_{s1} = 2\pi/1.5$, $k_{s2} = 2\pi/0.86$, $k_{s3} = 2\pi/0.80$, $k_{s4} = 2\pi/0.45$.

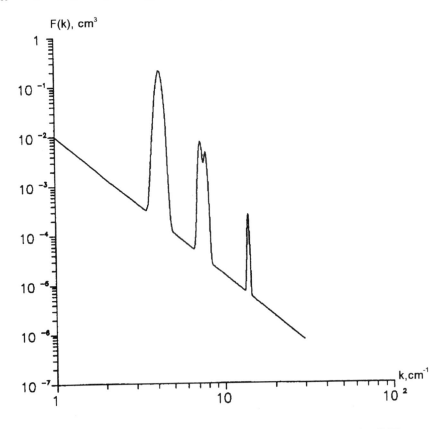

Figure 5.37. Modified spectrum of small-scale roughness on the ocean surface (5.22).

The calculations were made for a viewing angle of 75°. Figure 5.38 shows the experimental data of the spectral contrast in ocean surface brightness temperature variations on the eve of the origin of tropical cyclone *Warren*, as well as theoretical ones calculated via formula (5.20) using the modified spectrum of small-scale roughness (5.22). It can be seen that experimental and theoretical data are in good agreement qualitatively and quantitatively. Thus, by varying the parameters of a modified spectrum of small-scale roughness on the ocean surface (5.22), it is possible to achieve both positive and negative contrasts of ΔT_b at some frequencies (at least on vertical polarization) and to explain the negative correlation (see Figure 5.35) or non-correlation (see Figure 5.19 and Figure 5.31) of signals in the radiometer channels.

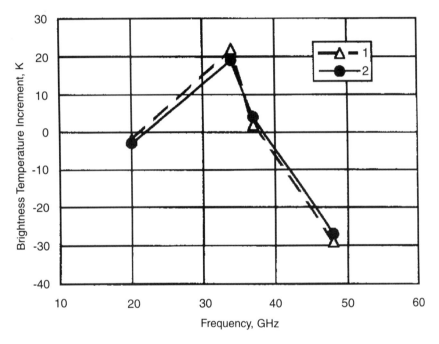

Figure 5.38. Spectral dependence of ocean surface brightness temperature variations on the eve of tropical cyclone *Warren*'s origin. Viewing angle 75°, vertical polarization. 1—experimental data; 2—calculated data by using modified spectrum of small-scale roughness on the ocean surface (5.22).

There is evidently a strong case that the transformation of the capillary wave spectrum is responsible for spectral variations in ocean surface brightness temperature on the eve of tropical cyclone *Warren*'s origin, as well as in Kuroshio region and 'relic rain'. In all cases mentioned, the intensive thermohaline processes of ocean active layer take place. Although, there is direct evidence of influencing the thermohaline processes onto the generation of surface capillary wave [66], so far the question of the mechanism of the action which could cause so significant a rearrangement of the capillary wave spectrum is still open. We note only that the cyclic frequencies of capillary waves lie in an acoustic range and consequently they can be raised by acoustic means, for example 'Faraday's ripple' [112, 113].

5.7 DIAGNOSTICS OF ANOMALOUS CYCLONE TRAJECTORY IN NORTH-WESTERN PACIFIC

The development of methods for remote diagnostics and forecasting of cyclone movement trajectories has great purpose for a disaster warning, particularly when typhoons or hurricanes drastically change direction. Forecasting the movement of tropical cyclones has been studied by many scientists [114–122]. Nevertheless, the forecasting of a cyclone's anomalous trajectory is still a difficult problem.

In 1990 the joint U.S.–Russian (Soviet) field experiments 'TCM-90' and 'Typhoon-90' were carried out in the North-Western Pacific to study tropical cyclone trajectories [123, 124]. The main goal of the Russian part of the 'Typhoon-90' experiment was to study the mechanisms resulting in anomalous tropical cyclone trajectories. There are three mechanisms that could be responsible for the anomalous trajectories of cyclone movement:

— action of environmental large-scale (background) flow;
— influence of the thermal properties of the upper ocean layer;
— change of asymmetry of thermodynamic structure of cyclone central zone.

The anomalous cyclone trajectory is defined as a change of forward movement direction of 45° or more or a change of its speed of 5 m/s or more during the next 6 hours with respect to its direction and speed during the previous 6 hours [123].

One of the factors determining the intensity and trajectory of a cyclone is the distribution of water temperature and heat content of the upper ocean layer in the region of the cyclone's existence. The influence of ocean surface temperature on the movement of tropical cyclones has been studied by many authors [125–130]. It has been shown that tropical cyclone trajectories to some extent are defined by the areas of warm and cold water, located at a distance of 300–500 km from their center [129, 130]. It was also shown [107, 127, 131] that typhoons frequently slow down or describe loops and change direction when they meet a sharp change in thermal non-homogeneities in ocean on the scale which is comparable to that of a cyclone active zone (which on the average occupies 200–400 km). These investigations are mainly made in tropical and subtropical regions. Typhoons often leave the middle latitudes and reach the Russian Far East, where their movement sometimes is of an anomalous type. Figure 5.39 shows the combined map of trajectories of typhoons during 1975–1981 in the North-Western Pacific, which have reached the Russian Far East [132]. The cyclone anomalous trajectories are observed in the Sea of Okhotsk and near the Kamchatka peninsula.

Below we consider the experimental results of airborne microwave studies of cyclone movement by an IL-18 research aircraft in the North-Western Pacific (sea of Okhotsk) during September, 1989. These experimental data provide additional evidence that the ocean thermic influence could be responsible for the anomalous character of its motion. By means of microwave mapping of the sea surface, the area of about 500 km size with

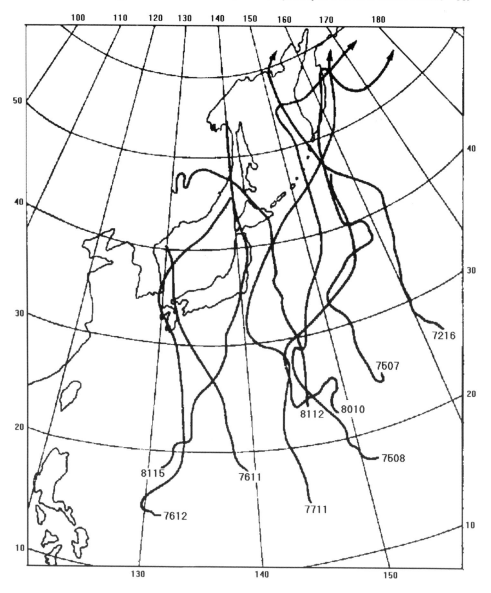

Figure 5.39. Typhoon trajectories reaching the Russian Far East over the period of
1975–1981 [132]

the anomalous spectral variations of brightness temperature has been detected in front of
the moving cyclone, where 7–8 hours later the cyclone has changed its trajectory [133].

The pattern of experiment is shown in Figure 5.40. The cyclone trajectory on
September 4–5, according to the data from the Japanese Meteorological Center, is presented in Figure 5.40. The cyclone was at the filling stage with a pressure in the center of

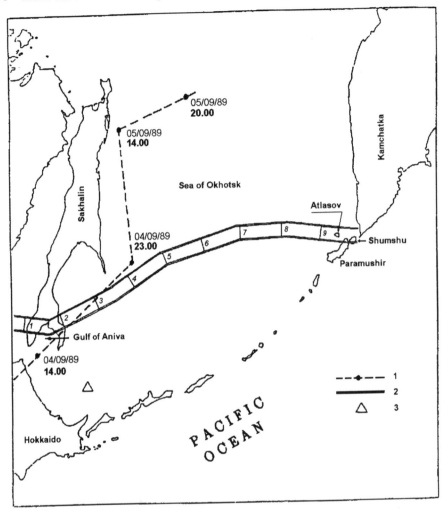

Figure 5.40. Pattern of airborne experiment on microwave studying the anomalous cyclone trajectory in the Sea of Okhotsk, September, 1989: 1—cyclone trajectory; 2—airborne microwave mapping region by IL-18; 3—hydrological station No. 1878 by R/V *Academician Nesmeyanov*.

1002–1005 mb. On September 4, at about 14.00 local solar time, the cyclone had arrived at the Sea of Okhotsk and its speed of forward movement exceeded 40 km/h. It moved in a northeast direction to a central part of the Sea of Okhotsk up to 23.00 on September 4, where it changed direction to the north. On this part the cyclone moved with a speed of about 30 km/h up to 14.00 on September 5, when it changed its forward movement again to the northeast.

Thus, the cyclone trajectory is a zigzag curve, which bends around the central part of the Sea of Okhotsk. The change in direction reaches 50–70°, which characterizes the cyclone trajectory as anomalous. A reduction in the cyclone's forward movement speed is also observed.

The airborne microwave mapping (section) of the Sea of Okhotsk was carried out by means of a microwave imaging radiometer [67] on September 4, between 14.22 and 16.15 and about 7–8 hours before as it has changed its trajectory. At the same time the cyclone 'eye' was located between Hokkaido and Sakhalin Islands (see Figure 5.40). The aircraft flight height was 8 km and the swath width was 50 km. The general mapping length was about 1200 km. The corresponding microwave images at frequencies of 34 and 75 GHz (both horizontal polarization) are presented in Figure 5.41 (colour section). They consist of separate frames. The position of each frame along the aircraft track is shown in Figure 5.40.

In the first frame (Figure 5.41, colour section), Sakhalin Island is observed. It is covered by cumulonimbus clouds of a cyclone's central part, about 100–150 km from its 'eye'. The microwave images allow us to distinguish the land–water boundary (Gulf of Aniva) only in the 34 GHz channel, while in the 75 GHz channel the upwelling radiation is defined mainly by the emission of clouds. Frame 2 is characterized by stratocumulus clouds; the next frames, 3 and 4, are characterized by a cloudy front with cumulonimbus clouds. It is clearly seen in the 75 GHz channel. Then, frames 5 to 9 are characterized by stratocumulus clouds.

Of special interest are the spectral variations of upwelling microwave emission detected in the central part of the Sea of Okhotsk. So, for example, there is an area of about 500 km of positive brightness temperature contrast of 30–40 K in the 34 GHz channel (frames 4–7 in Figure 5.41, colour section). At the same time, in the 75 GHz channel the fragments of structures of 50 to 100 km size are observed. They are characterized by both positive and negative brightness temperature contrast with respect to the background level (frame 8). The positive brightness temperature contrast reaches 12 K, while the negative reaches 18 K. The microwave mapping of the land surface obtained under the same atmosphere conditions is presented in frame 9. In this case, the land–water boundaries of Atlasov, Shumshu and Paramushire Islands are clearly seen on microwave images in both channels.

To classify the brightness temperature variation during airborne microwave mapping of the Sea of Okhotsk, the notion of interchannel derivative was used. We found that dT_b^{75H}/dT_b^{34H} is positive everywhere and equals a limited value; but, only in the area corresponding to frames 4–7, we have two mutually exclusive situations $(dT_b^{75H}/dT_b^{34H}) \to \infty$, and $(dT_b^{75H}/dT_b^{34H}) \to 0$. The area of such anomalous variations in microwave emission is propagated up to 500 km. According to the data analysis, presented in the previous sections, we may conclude that this microwave phenomenon cannot be related to the meteorological variables of the atmosphere, but it was most probably caused by the oceanic processes.

Let us consider the hydrological conditions in this region. The vertical profiles of temperature, salinity and density measured by CTD-sonde on September 10, 1989 are shown in Figure 5.42. Hydrological measurements were carried out during the 16th voyage of research vessel *Academician Nesmeyanov* (st. 1878) near to 45°05.7′N, 145°07.8′E, not far from the cyclone's track (150 km) and 6 days after the cyclone's passing (see Figure 5.40). The *in situ* data (see Figure 5.42) present a non-typical pattern

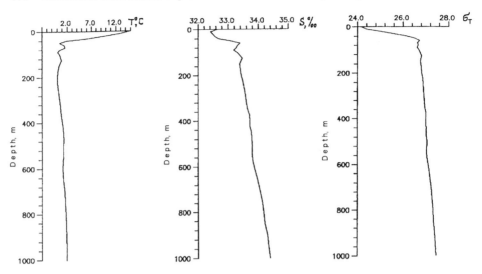

Figure 5.42. Vertical profile of temperature T, salinity S, and relative density σ_T at station No. 1878 of R/V *Academician Nesmeyanov*, on September 10, 1989. Sea of Okhotsk, 45°05.7′N, 145°07.8′E

for the August–September period. There is a thin, warm water layer near the surface, where the temperature is rapidly reduced from 15°C on the surface to 1.0–2.0°C at a depth of 30–40 m. Usually, the near-surface mixing layer of 20–25 m depth is presented due to wind wave propagation [134]. Then the temperature is reduced from 1.0 to 2.0°C at a depth of 40 m up to 0°C at a depth of 200 m, and then it slightly increases with depth. The upper layer up to 180 m is characterized by a thermohaline structure, where inversions of temperature and salinity are observed. The inversions of the density profile indicate that the hydrostatic instability takes place there.

The ocean's influence on the cyclone's trajectory is defined not only by the ocean surface temperature distribution, but by the thermal non-homogeneities of the ocean active layer. It was shown [131] that cyclone interaction with the ocean is propagated up to a depth of 1000 m. Figure 5.43 shows the experimental results of the reaction of the South China Sea on tropical storm *Maimy* (maximum wind speed was 22 m/s). The temperature difference between the water layer and the depth before and after cyclone passage indicates that the cooled water anomalies caused by the cyclone's influence are located at a depth of up to 800 m and more. The maximum water cooling of the upper ocean layer is observed in the zone of radius of maximum winds. According to [125–128, 131], the main mechanism responsible for the water cooling is upwelling. Its contribution to water cooling comprises about 58%, while the turbulence implication provides about 30%, and the energy exchange between the ocean and atmosphere comprises only 12% of this contribution.

In our airborne experiment in the Sea of Okhotsk, the detected area of 500 km size with the anomalous variation of microwave emission in the front of the moving cyclone indicates the development of instability of the thermohaline processes in the ocean active layer. This conclusion can be reached by using the results and analysis described in

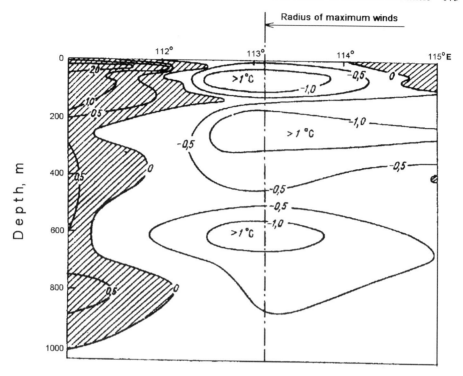

Figure 5.43. Reaction of South China Sea to tropical storm *Maimy* (8823). Temperature difference between water layer before and after cyclone passage (latitude section). The cooling of water is observed up to 1000 m [131].

sections 5.4 and 5.5. The instability of processes in the ocean active layer could be caused by the influence of the cyclone itself, which ultimately has resulted in the origin of the upwelling. This, in turn, would be responsible for the change in direction of the cyclone movement.

5.8 OBSERVATIONS OF OCEANIC PROCESSES IN NORTHERN ATLANTIC BY USING DMSP SSM/I DATA

In this section we discuss the applications of satellite microwave data provided by the SSM/I instrument for observations of oceanic processes similar to those described in previous sections, when a shipboard and airborne multichannel microwave radiometer was used. Our analysis of the SSM/I data is restricted by a period through January 1997 and a region of the Northern Atlantic where the same microwave phenomenon has been detected.

SSM/I data were provided by the Global Hydrology Resource Center at the Global Hydrology and Climate Center, NASA Marshall Space Flight Center, Huntsville, Alabama. SSM/I performance characteristics and orbit parameters for DMSP spacecraft are described in section 4.5.

The source of the GHRC's SSM/I data is the Fleet Numerical Meteorology and Oceanography Center (FNMOC). FNMOC is the Department of Defense's (DoD) center of expertise for DMSP passive microwave data processing. The DoD has an agreement with the National Oceanographic and Atmospheric Administration (NOAA) under which they share data via the Shared Processing Network (SPN). FNMOC generates SSM/I antenna temperature files known as Temperature Data Records (TDRs) and sends them to NOAA/NESDIS. NESDIS performs minimal file unpacking and makes the data available to the GHRC. Files are usually available within 4–6 hours of data acquisition by FNMOC.

Each day the GHRC performs a four-step quality control process on the previous day's data. These steps are: date stamping; file merging; navigation checking; and calibration checking. After these quality control steps, data are read for each pass (ascending or descending), sequentially for an entire day. Antenna temperatures are then converted to brightness temperatures by using the Remote Sensing System's antenna pattern correction (APC) and along-scan bias corrections [135, 136]. Each day, full resolution brightness temperatures and reduced resolution 'gridded' data sets for seven channels and both ascending and descending passes are generated.

Figure 5.44 (colour section) shows the global gridded SSM/I data for the brightness temperature at a frequency of 19 GHz, horizontal polarization for descending passes of DMSP F13 spacecraft on January 10, 1997. Figure 5.45 (colour section) shows the same data for the brightness temperature, but at a frequency of 19 GHz vertical polarization and for ascending passes. Both ascending and descending passes result in 24 hours global coverage for one satellite. Only the diamond-shaped areas are missed near the equator, but these regions are covered after 72 hours.

The SSM/I measurements provide many environmental products: ocean surface wind speed [9–11]; integrated water vapor [12]; cloud liquid water content [13]; precipitation; ice area covered and age; snow water content [15], as well as, ocean surface wind direction [11].

GHRC produces SSM/I geophysical products, by using Wentz's SSM/I Benchmark Pathfinder Algorithm [11, 136]. From the SSM/I brightness temperature swath data of the DMSP F13 and DMSP F14 satellites, the GHRC generates three geophysical products at swath and gridded resolution: integrated water vapor (IWV); cloud liquid water (CLW); and oceanic wind speed (OWS).

The algorithms for retrieving the integrated water vapor and cloud liquid water in the atmosphere use the SSM/I 22 GHz vertically polarized channel and the dual-polarized 37 GHz channels [11]. The valid ranges for determining integrated water vapor and cloud liquid water are $0–10\,g/cm^2$ and $0–1000\,mg/cm^2$, respectively. Figure 5.46 and Figure 5.47 (both in colour section) show these geophysical products from the DMSP F13 satellite at a gridded resolution on January 10, 1997 for descending passes.

The algorithm for computing the oceanic wind speed uses only the SSM/I dual-polarized 37 GHz channels [11]. The valid range for determining the oceanic wind speed is $0–40\,m/s$. Oceanic wind speed cannot be retrieved when the CLW is greater than $18.3\,mg/cm^2$. Figure 5.48 (colour section) shows the oceanic wind speed from the DMSP F-13 satellite at gridded resolution on January 10, 1997 for descending passes.

Next we consider the upwelling microwave radiation of the ocean–atmosphere system at frequencies of 37 and 85 GHz for vertical and horizontal polarization on January 10, 1997, to analyze the character of brightness temperature variations due to both meteorological variables of the atmosphere and the oceanic processes directly. Figure 5.49 (colour section) shows the microwave images of the Northern Atlantic for channels 37H, 37V, 85H and 85V. We picked out two areas, A and B, for analysis (see Figure 5.49, colour section). Area A occupies a northeastern part of the Atlantic Ocean and a southern part of the Norwegian Sea and is characterized by intensive variations in integrated water vapor from 0.5 to 3 g/cm^2 (Figure 5.46, colour section), cloud liquid water from 10 to 120 mg/cm^2 (Figure 5.47, colour section), and near-surface wind speed from 8 to 22 m/s (Figure 5.48, colour section).

Area B occupies a northern part of the Norwegian Sea and is characterized by very small variations in integrated water vapor and cloud liquid water, because the amounts are less than 0.5 g/cm^2 (Figure 5.46, colour section) and 10 mg/cm^2 (Figure 5.47, colour section), respectively, but the near-surface wind speed is changed within 20–25 m/s (Figure 5.48, colour section).

To analyze the brightness temperature variations in the areas A and B the regression diagrams are used. A high positive correlation is demonstrated between brightness temperature variations in area A for a combination of two channels, 37V and 37H (see Figure 5.50), and 85V and 85H (see Figure 5.51). In this case, the meteorological variables—humidity, cloud liquid water content, wind speed, as well as air and water surface temperature—are mainly responsible for brightness temperature changes. The maximum total brightness temperature contrasts corresponding to variations in these meteorological variables are: $\Delta T_b^{37H} \approx 95$ K, $\Delta T_b^{37V} \approx 46$ K, $\Delta T_b^{85H} \approx 74$ K, $\Delta T_b^{85V} \approx 29$ K. Precipitation is excluded from consideration. The algorithm for detecting rain is described, for instance, in the literature [11, 137].

The linear dependence between small brightness temperature increments for a combination of these two pairs of channels for area A can be described according to (5.15) and Figure 5.50 and Figure 5.51 as follows:

$$\Delta T_b^{37V} = \left(\frac{dT_b^{37V}}{dT_b^{37H}} \right)_A \Delta T_b^{37H}$$

and (5.23)

$$\Delta T_b^{85V} = \left(\frac{dT_b^{85V}}{dT_b^{85H}} \right)_A \Delta T_b^{85H},$$

where $\left(dT_b^{37V} / dT_b^{37H} \right)_A \approx 0.5$; and $\left(dT_b^{85V} / dT_b^{85H} \right)_A \approx 0.4$, the interchannel derivative.

As to area B, a high positive correlation is demonstrated only between brightness temperature variations for the two-channel combination of 37V and 37H (see Figure 5.50), while the correlation dependence between brightness temperature variations for the

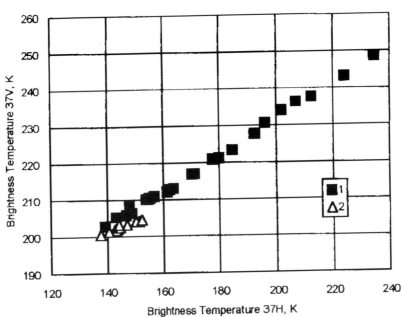

Figure 5.50. Regression dependence between SSM/I brightness temperature variations in the channels of 37 GHz horizontal polarization and 37 GHz vertical polarization in accordance with marked areas A and B in Figure 5.49: 1—area A; 2—area B.

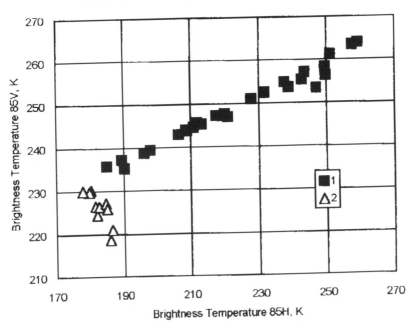

Figure 5.51. Same as Figure 5.50, but for 85 GHz horizontal polarization and 85 GHz vertical polarization.

two-channel combination of 85V and 85H (see Figure 5.51) has negative sign. The maximum contrasts in brightness temperature variations being observed in area B are: $\Delta T_b^{37H} \approx 15$ K, $\Delta T_b^{37V} \approx 7$ K, $\Delta T_b^{85H} \approx 10$ K, $\Delta T_b^{85V} \approx -12$ K. We can see that brightness variations in the 85V channel have negative contrast. The interchannel derivative in area B for the two-channel combination of 37V and 37H is $\left(dT_b^{37V} / dT_b^{37H} \right)_B \approx 0.5$, and

for the combination of channels 85V and 85H is $\left(dT_b^{85V} / dT_b^{85H} \right)_B \approx -1.2$.

Let us consider the microwave images of the Norwegian Sea in more detail to analyze the spatial and temporal structure of the microwave phenomenon, where the interchannel derivative is

$$\left(\frac{dT_b^{85V}}{dT_b^{85H}} \right) < 0. \tag{5.24}$$

Figure 5.52 (colour section) shows the microwave images of the Norwegian Sea from January 8 to January 12, 1997, at frequencies of 37 GHz and 85 GHz, both vertical polarization. Since January 10, the satellite microwave data for both ascending and descending passes of DMSP F13 are shown in Figure 5.52, so the temporal interval for observing the Norwegian Sea is about 12 hours. We analyze the microwave images of the sea surface for two channels, 37V and 85V, to underline the spectral peculiarity of this microwave phenomenon, because it is seen only at a frequency of 85 GHz vertical polarization.

The said microwave phenomenon where the expression (5.24) is true appeared on January 10 (Figure 5.52(b)), whereas it was absent the day before (Figure 5.52(a)). Its lifetime is approximately 2 days. The microwave structure is located between 64.5° and 70.5°N latitude, and 13.0° and 19.5°E longitude. Its spatial size reaches about 650 km in length and can be related to the processes of synoptic scale. This microwave phenomenon is clearly observed during January 10 and 11 (Figures 5.52(b), (c), (d)); then it begins to be destroyed (Figure 5.52(e)), and finally disappears on January 12 (Figure 5.52(f)). The maximum negative brightness temperature contrast reaches $\Delta T_b^{85V} \approx 12$–14 K.

The analysis shows that this microwave anomaly is not related to the meteorological variables of the atmosphere, but is most probably due directly to the sea surface factors. In addition, the phenomenon observed in the Norwegian Sea by means of a satellite microwave radiometer is similar to those detected by means of shipboard and airborne microwave radiometers in the Kuroshio region (see sections 5.4 and 5.5) and South China Sea (see section 5.3). It is significant that the spectral and polarization microwave effect in the Norwegian Sea described by (5.24) is detected under storm conditions of sea surface, when the wind speed is 20–25 m/s (see Figure 5.48, colour section).

Now, we discuss the hydrological situation in the Norwegian Sea to determine the oceanic processes, which could be responsible for the said microwave phenomenon. In winter, the highest sea surface temperature of 6–7°C is observed in the southern part of the Norwegian Sea; in the northern part this is 2–3°C. In summer, this sea surface temperature corresponds to 12–13°C and 5–7°C respectively. The water temperature smoothly decreases with a depth being positive to horizons about 1000 m [134].

In winter, the sea surface salinity equals 34–35‰ and its distribution is quite monotonic. Isohaline 34‰ runs along the Scandinavian peninsula coast. To the west from ones and in all space the surface salinity is 35‰. The salinity slightly increases to 35.05–35.10‰ with a depth to 500 m horizon. The summer salinity distribution is weakly distinguishable from the winter one [134].

The Norwegian Sea is characterized by the sea climate of temperate latitudes. The air temperature over the sea in winter time has mean monthly values of from –4°C in the north to +4°C in the south. The zero's isotherms of January runs from Iceland to Bear Island. The mean wind speed in the winter is 8–10 m/s and 5–6 m/s in the summer [134].

It is just during the winter that the active ocean layer undergoes intensive restructuring. Figure 5.53 shows the typical seasonal evolution of the vertical temperature profile for moderate latitude of the Northern Atlantic [138]. At the end of winter, the thermocline is disposed at a depth of 150 m as a result of convective mixing and the influence of storms. In the summer period the accumulation of heat in the upper ocean layer results in the thermocline with a depth of 30–60 m. At the beginning of the summer, the thermocline may yet be destroyed by the strong winds, but at the end of summer, the vertical water temperature gradient is that the strongest winds cannot transfer enough mechanical energy to the ocean to destroy a thermocline. At this stage, which continues up to the beginning of the winter, a thermocline is rapidly strengthened due to energy accumulation in the thin layer above a seasonal thermocline. According to [138], this typical

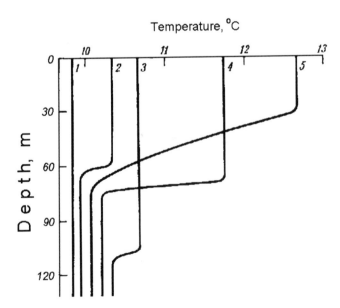

Figure 5.53. Typical vertical distribution of water temperature for Northern Atlantic [138]:
1—end of winter; 2—springtime; 3—summer; 4—autumn; 5–beginning of winter.

seasonal evolution of the vertical temperature profile could be distorted by the oceanic currents.

One of the main features of North European basin circulation is the perpetual inflow of Atlantic waters through the Faroe–Shetland strait forming the Norwegian Current extending northeastward from the warm North Atlantic Current, contributing to maintaining the Norwegian Sea free of ice [139] (Figure 5.54). This current gives two branches in the Barents Sea (Nordcap Current) and in the Arctic Basin (West-Spitsbergen Current). Part of the West-Spitsbergen Current waters comes back to the south as the East-Greenland Current, taking its own origin in the Arctic Basin. In the northern part of the Norwegian Sea there is the cool Bear Island Current, coming from the Barents Sea. The complicated circulation pattern in the surface waters of this region is characterized as the Norwegian energy active zone [140, 141].

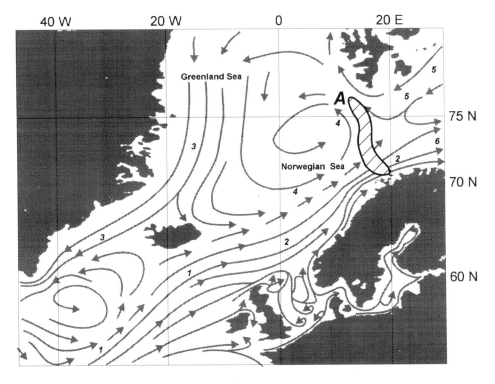

Figure 5.54. Large-scale surface currents of Northeastern Atlantic [139] and position of microwave anomaly (A) detected by SSM/I on January 10–11, 1997, according to Figure 5.52: 1—North Atlantic Current; 2—Norwegian Current; 3—East-Greenland Current; 4—West-Spitsbergen Current; 5—Bear Island Current; 6—Nordcap Current.

The North European basin has been intensively studied by Russian oceanographers. During 1976–1985 a number of research vessel missions to the Norwegian Sea and Greenland Sea have been organized in accordance with the programs of 'POLEKS-SEVER' and 'SECTIONS'. The investigations were carried out by the Research Institute of Arctic and Antarctic, Saint-Petersburg, and their results have been exhaustively presented in a monograph [140].

One of the main factors in forming the variability of the ocean active layer in the Norwegian energy active zone is the energy exchange with the atmosphere. According to analysis of multiannual readings of heat balance [140] it has been recognized that during the winter period the energy sources in the atmosphere arise on the warm currents. The most intensive source arises on the West-Spitsbergen Current. The calculations of energy exchange between ocean and atmosphere based on the daily meteorological observations in the winter have shown that there are intensive synoptic splashes of fluxes from ocean atmosphere. Figure 5.55 shows the fluxes of sensible heat and latent heat from ocean to atmosphere in the region of West-Spitsbergen Current and anticyclonic water circulation near 70.0°N, and 3.5°E during October, 1986. It illustrates the 'impulse' character of synoptic disturbances of energy exchange between ocean and atmosphere. The most intensive heat transfer is observed during October 7–9, where there are fluxes of both sensible heat and latent heat. Some feeble pulses of latent heat flux are observed during October 12–15, while the sensible heat flux remains on the background level. It can be seen that the splashes of fluxes of sensible heat and latent heat from ocean to atmosphere exceed their background meaning more than one order. The period of anomalous fluctuations of energy exchange between ocean and atmosphere in Norwegian energy active zone is equal to 1–3 days. It correlates with the lifetime of the said microwave anomaly, which equals about 2 days. On the other hand, the relation between ocean surface microwave emission and sensible and latent heat fluxes at the interface during the evolution of

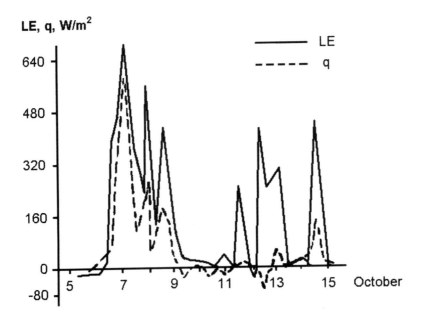

Figure 5.55. Turbulence fluxes of sensible heat (q) and latent heat (LE) from ocean to atmosphere in the Norwegian Sea near 70.0°N, 3.5°E, during October, 1986 [140].

thermal state of the ocean–atmosphere system has been shown in the literature [142, 143].

The above-mentioned microwave anomaly is located in the northeastern part of the Norwegian Sea, crossing the Norwegian Current and propagating along the West-Spitsbergen Current (Figure 5.54). In this region of the Norwegian Sea a polar front is located [140, 141, 144]. According to [144], the position of the polar front at a depth of 200 m here is characterized by less seasonal variability than one on the surface. This in turns allows us to relate the said sea surface microwave effect to the external hydrological conditions at the depth.

The main conclusion from the results discussed is that the spectral and polarization brightness temperature variations of ocean surface, which are related to oceanic processes, can be detected by means of an SSM/I satellite microwave radiometer. We were going to demonstrate the possibility of satellite passive microwave means of observing the processes of the ocean active layer, but not to carry out its monitoring. In our case, only the frequency channel of 85 GHz vertical polarization provided useful information, and the question naturally arises about the probability of detecting these processes by means of an SSM/I, particularly, and microwave radiometry at all. Because, according to (5.11), the said microwave phenomenon can be characterized by a sufficiently high spectral selectivity, less than 7%.

Summarizing all that has been considered here regarding experimental results of shipboard, airborne, and satellite observations of the said microwave phenomenon we can enumerate a number of operating frequencies regarding its detection. These are: 34, 37, 48, 75 and 85 GHz. On the other hand, this effect has not been detected at the following frequencies used for remote sensing: 19, 20 and 22 GHz.

The analysis shows that the most effective frequency band for detecting the spectral variations in ocean surface microwave emission, which are a consequence of the manifestation of the deep oceanic processes, ranges approximately from 30 to 90 GHz. This fact was taken into account in the consideration of the operating frequencies of the 'Meteor-3M' MTVZA satellite microwave radiometer (see section 4.4). A number of its operating frequencies include some non-typical ones, especially for oceanographic research: 33, 36.5, 42, 48 and 91.6 GHz.

REFERENCES

[1] Ulaby, F.T., Moor, R.K., and Fung, A.K. (1986) *Microwave Remote Sensing: Active and Passive*, Vol. 1, N.Y.: Artech House 1981, and Vol. 3, N.Y.: Artech House.

[2] Kondratyev, K.Ya., Melentyev, V.V., and Nazarkin, V.A. (1992) Remote sensing of waterareas and watersheads (microwave methods). Saint-Petersburg: Gidrometeoizdat. 248 pp. (in Russian).

[3] Stepanenko, V.D., Shukin, G.G., Bobylev, L.P. and Matrosov, S.Yu. (1987) Radio-thermosensing in meteorology. Leningrad: Gidrometeoizdat, 284 pp. (in Russian).

[4] Kondratyev, K.Ya. and Timofeev, Yu.M. (1978) Meteorological remote sensing of atmosphere from space. Leningrad: Gidrometeoizdat, 280 pp. (in Russian).

[6]   Shutko, A.M. (1986) Microwave radiometry of water surface and ground. Moscow: Nauka, 189 pp. (in Russian).

[7]   Timofeev, Yu.M., Polyakov, A.V., Vasilev, A.V., Shul'gina, E.M. and McClatchy, M. (1997). Microwave temperature–humidity probing of atmosphere from space, *Atmospheric and Oceanic Physics*, **33**, No. 1, 53–61. Izvestiya AN, (in Russian).

[8]   Holinger, J.P., Pierce, J.L. and Poe, G.A. (1990) SSM/I instrument and evaluation, *IEEE Trans. Geosci. Remote Sensing*, **28**, No. 5, 781–790.

[9]   Wentz, F.J., Mattox, L.A. and Peteherych, S. (1986) New algorithm for microwave measurements of ocean winds: Applications to SEASAT and Special Sensor Microwave/Imager, *J. Geophys. Res.* **91**, C2, 2279–2307.

[10] Goodberlet, M.A., Swift, C.T. and Wilkerson (1990) Ocean surface wind speed measurements of the special sensor microwave/imager (SSM/I), *IEEE Trans. Geosci. Remote Sensing*, **28**, No. 5, 823–828.

[11] Wentz, F.J. (1992) Measurement of oceanic wind vector using satellite microwave radiometer, *IEEE. Trans. Geosci. Remote Sensing*, **30**, No. 5, 960–972.

[12] Alishous, J.C., Snyder, S.A., Vongsathorn, J. and Ferraro, R.R. (1990) Determination of oceanic precipitable water from the SSM/I, *IEEE Trans. Geosci. Remote Sensing*, **28**, No. 5, 811–816.

[13] Alishouse, J.C., Snider, J.B., Westwater, E.R., Swift, C.T., Ruf, C.S., Snyder, S.A., Vongsathorn, J. and Ferraro, R.R. (1990) Determination of cloud liquid water content using the SSM/I, *IEEE Trans. Geosci. Remote Sensing*, **28**, No. 5, 817–822.

[14] Parkinson, C.L., Comiso, J.C., Zwally, H.J., Cavaliery, D.J., Gloersen, P. and Campbell, W.J. (1987) Arctic sea ice, 1973–1976: satellite passive-microwave observations, *II NASA SP-489. W. DC.* 296 pp.

[15] Ferraro, R.R., Weng, F., Grody, N.A. and Basist, A. (1996) An eight-year (1987–1994) time series of rainfall, clouds, water vapor, snow cover, and sea ice derived from SSM/I measurements, *Bull. Amer. Meteor. Soc.*, **77**, 891–905.

[16] Kunkee, D.B. and Gasiewski, A.J. (1997) Simulation of passive microwave wind direction signatures over the ocean using an asymmetric-wave geometrical optics model, *Radio Science*, Vol. 1, 59–78.

[17] Bates, J.J. (1991) High-frequency variability of special sensor microwave imager derived wind speed and moisture during an interseasonal oscillation, *J. Geophys. Res., Supplement*, **96**, 3411–3423.

[18] Rasmusson, E.M. and Carpenter, T.H. (1982) Variations in tropical sea surface temperature and surface wind fields associated with Southern Oscillation/El-Nino, *Mon. Wea. Rev.*, **110**, 354–384.

[19] Atlas, D., Beal, R.C., Brown, R.A., Mery, P.D., Moore, R.K., Rapley, C.G. and Swift, C.T. (1986) Problems and future directions in remote sensing of the oceans and atmosphere: a workshop report, *J. Geophys. Res.*, **91**, No. C2, 2525–2548.

[20] Gus'kov, G.Ya., Moiseev, S.S. and Cherny, I.V. (1991) Secondary instabilities in the ocean–atmosphere system and microwave diagnostics of natural calamities, *Preprint IKI RAN (Space Research Institute)*, N1762, Moscow, 34 pp.

[21] Ostrovsky, L.A., Rybak, S.A. and Tzimring, L.S. (1986) Negative waves in hydrodynamics, *Uspekhi Fysicheskih Nauk*, **150**, No. 3, 417–437 (in Russian).

[22] *Nonlinear Waves. Self-Organization.* (Eds. Gaponov-Grekhov, A.V. and Rabinovich, M.I.), Moscow: Nauka, 1983. (in Russian).

[23] Fedorov, K.N. (1976) Ocean thermohaline fine structure. Leningrad: Gidrometeoizdat, 183 pp. (in Russian).

[24] Fedorov, K.N. (1983) Physical nature of structure of oceanic fronts. Leningrad: Gidrometeoizdat, 296 pp. (in Russian).

[25] Moiseev, S.S. and Sagdeev, R.Z. (1986) The problems of secondary instabilities in hydrodynamics and plasma, *Izvestiya Vuzov—Radiofyzika*, **29**, No. 9, 1067–1072. (in Russian).

[26] *Nonlinear Waves. Structures and Bifurcation.* (Eds. Gaponov-Grekhov, A.V. and Rabinovich, M.I.), Moscow: Nauka, 1987, 398 pp. (in Russian).

[27] Veselov, V.M., Gerbek, E.E., Zabyshny, A.I., *et al.* (1989) On the examination of physical model of the origin of large-scale vortexes with non-zero helicity. *Preprint IKI RAN (Space Research Institute)*, No. 1604, Moscow, 12 pp. (in Russian).

[28] Seidov, D.G. (1989) *Synergism of Oceanic Processes.* Leningrad: Gidrometeoizdat, 288 pp. (in Russian).

[29] Russian Federation Patent No. 2,047,874.

[30] United States Patent No. 5,631,414.

[31] Apel, `I.R., Byrne, H.M., Proni, S.R. and Charnell, R.L. (1975) Observations of oceanic internal and surface waves from ERTS, *J. Geophys. Res.*, **80**, No. 6, 865–881.

[32] Brown, W.E., Elachi, C. and Thompson, I.W. (1976) Radar imaging of ocean surface patterns, *J. Geophys. Res.*, **81**, No. 15, 2657–2667.

[33] Hughes, B.A. and Grant, H.L. (1978) The effect of internal waves on surface wind waves. I. Experimental measurements; Hughes, B.A. The effect of internal waves on surface wind waves. 2. Theoretical analysis, *J. Geophys. Res.*, **83**, No. C1, 443–454, 455–469.

[34] Bravo-Zhyvotovsky, D.M., Volodina, N.I., Gordeev, A.B., *et al.* (1982). Studies of influencing oceanic internal waves onto surface waves by remote means, *Dokl. Akad. Nauk SSSR*, **265**, No. 2, 457–460 (in Russian).

[35] Veselov, V.M., Davydov, A.A., Skachkov, V.A., Cherny, I.V and Volyak, K.I. (1984) Shipboard remote microwave measurements of internal waves, *Physics of Atmosphere and Ocean*, **20**, No. 3, 308–317, Izvestiya AN ASSR, (in Russian).

[36] Thompson, D.R. and Jensen, J.R. (1993) Synthetic aperture radar interferometry applied to ship-generated internal waves in the 1979 Loch Linnhe experiment, *J. Geophys. Res.*, **98**, No. 10, 259–269.

[37] Brandt, P., Alpers, W., and Backhaus, J.O. (1996) Study of the generation and propagation of internal waves in the Strait of Gibraltar using a numerical model and synthetic aperture radar images of the European ERS-1 satellite, *J. Geophys. Res.*, **101**, No. 14, 237–252.

[38] Cherny, I.V. (1982) MM-wave radiometer–scatterometer for remote sensing of

sea surface, Preprint IKI RAN (Space Research Institute) N. 689, Moscow, 19 pp. (in Russian).

[39] Konyayev, K.V. and Sabinin, K.D. (1973) New data on internal waves in the ocean obtained with distributed temperature sensors. *Dokl. Akad. Nauk SSSR*, **209**, No. 1, 86–89 (in Russian).

[40] Benjamin, T.B. (1967) Internal waves of permanent form in fluids of great depth, *J. Fluid Mech.*, **29**, Pt. 3, 559–579.

[41] Whitham, G. (1974) *Linear and Nonlinear Waves*. Wiley.

[42] Osborne, A.R. and Burch, T.L. (1980. Internal solitons in the Andaman Sea, *Science*, Vol. 208, No. 4443, pp. 451–460.

[43] Monin, A.S. (1978) *Oceanology. Hydrodynamics of Ocean*. Vol. 2, Moscow: Nauka, 456 pp. (in Russian).

[44] Basovich, A.Ya. and Talanov, V.I. (1977) On transformation of spectrum of short surface waves on the non-homogeneous currents, *Physics of Atmosphere and Ocean*, **13**, No. 7, 766 (in Russian).

[45] Basovich, A.Y. (1979) Transformation of spectrum of surface waves under influence of internal waves, *Physics of Atmosphere and Ocean*, **15**, No. 6, Izvestiya AN SSSR, 655. (in Russian).

[46] Miropolsky, Yu.Z. (1981) *Dynamics of Internal Waves in the Ocean*. Leningrad: Gidrometeoisdat, 302 pp. (in Russian).

[47] Peters, A.S. and Stoker, J.J. (1960) *Commun. Pure Applied Mathematics*, **13**, 115–124.

[48] Gasparovic, R.F., Chapman, R.D., Monaldo, F.M., Porter, D.L. and Sterner, R.E. (1993) Joint U.S./Russia Internal Wave Remote Sensing Equipment. Interim results. The Johns Hopkins University, Applied Physics Laboratory, USA, JHU/APL S1R-93U-011.

[49] Gasparovic, R.F. and Etkin, V.S. (1994) An overview of the Joint US/Russia internal wave remote sensing experiment. *IGARSS'94 Proceedings*, Pasadena, California, USA, Vol. II, pp. 741–743.

[50] Etkin, V.S., Trokhimovski, Yu.G., Yakovlev, V.V., and Gasparovic, R.F. (1994) Comparison analysis of Ku-band SLAR sea surface images at VV and HH polarization obtained during the Joint US/Russia Internal Wave Remote Sensing Experiment, *IGARSS'94 Proceedings*, Pasadena, California, USA, Vol. II, pp. 744–746.

[51] Ginsburg, A.I., Zatsepin, A.G., Sklyarov, and Fedorov, K.N. (1980) Precipitation effect on sub-surface ocean layer. *Okeanologiya*, **20**, No. 5, 828–936 (in Russian).

[52] Evans, M. (1971) Surface salinity and temperature "Signature" in the Northeastern Pacific. *J. Geophys. Res.*, **76**, No. 15, 3456–3461.

[53] Cherny, I.V. (1992) The "Relic rain" effect on sea surface microwave emission. *Proceedings of IGARSS'92 Symposium*, Houston, Texas, *IEEE 92CH3041-1*, Vol. I, pp. 254–256.

[54] Wu, S.T. and Fung, A.K. (1972) A non-coherent model for microwave emission and backscattering from the sea surface. *J. Geophys. Res.*, **77**, 5917–5929.

[55] Vilheit, T.T. and Chang, A.T.C. (1980) An algorithm for retrieval of ocean

surface and atmosphere parameters from observations of the scanning multi-channel microwave radiometer, *Radio Science*, **15**, 525–544.

[56] Wentz, F.J. (1983) A model function for ocean microwave brightness temperatures, *J. Geophys. Res.*, **88**, C3, 1892–1908.

[57] Fedorov, K.N. and Ginsburg, A.I. (1988) *Sub-surface Layer of the Ocean*. Leningrad: Gidrometeoizdat, 304 pp. (in Russian).

[58] Turner, J.S. (1973) *Buoyancy Effects in Fluids*. Cambridge University Press, 367 pp.

[59] Turner, J.S. (1978) Double-diffusive intrusions into a density gradient. *J. Geophys. Res.*, **83**, No. C6, 2887–2901.

[60] McDougall, T.J. (1985) Double-diffusive interleaving. Part I: Linear stability analysis. *J. Phys. Oceanogr.*, **15**, No. 11, 1532–1541.

[61] McDougall, T.J. (1985) Double-diffusive interleaving. Part II: Finite amplitude, steady state interleaving. *J. Phys. Oceanogr.*, **15**, No. 11, 1542–1556.

[62] Toole, J.M. and Georgi, D.T. (1981) On the dynamics and effects of double-diffusively driven intrusions. *Progress in Oceanography*, Vol. 10. Pergamon, pp. 123–145.

[63] Garret, C. (1982) On the parameterization of diapycnal fluxes due to double-diffusive intrusions. *J. Phys. Oceanogr.*, **12**, 954–959.

[64] Ruddick, B.R. and Turner, J.S. (1979) The vertical length scale of double-diffusive intrusions, *Deep-Sea Res.*, **26**, 903–913.

[65] Double-diffusive convection. Eds Brandt, A. and Fernando, H., *AGU Geophysical Monograph*, No. 94, 1996, 334 pp.

[66] Cherny I.V. (1985) Double-diffusive convection effect on capillary–gravity waves spectrum, *Dokl. Akad. Nauk SSSR*, **282**, No. 5, 1117–1120. (in Russian).

[67] Gorobetz, N.N., Zabyshny, A.I., Il'gasov, P.A., Cherny, I.V. and Sharapov, A.N. (1989) Multichannel microwave radiometer for remote sensing of ocean and atmosphere. *Preprint IKI RAN* (Space Research Institute) No. 1545, Moscow, 32 pp. (in Russian).

[68] Holinger, J.P. (1971) Passive microwave measurements of sea surface roughness. *IEEE Trans. Geosci. Electron.*, **9**, No. 3, 165–169.

[69] Keller, W.C. and Right, J.W. (1975) Microwave scattering and the straining of wind-generated waves, *Radio Science*, **10**, 139–147.

[70] Kawai, S. (1973) Generation of initial wavelets by instability of a coupled shear flow and their evolution to wind waves. *J. Fluid Mech.*, **93**, Pt. 4, 661–703.

[71] Ruvinsky, K.D. and Freidman, G.I. (1987) Fine structure of strong gravity–capillary waves. *Nonlinear Waves. Structures and Bifurcations*. (Eds Gaponov-Grekhov, A.V. and Rabinovich, M.I.), Nauka, Moscow, pp. 304–326 (in Russian).

[72] Monin, A.C. and Krasnitsky, V.P. (1985) *Phenomena on the Ocean Surface*. Leningrad: Gidrometeoizdat, 375 pp. (in Russian).

[73] Kravtsov, Yu.A. and Etkin, V.S. (1983) Wind waves as autooscillation process, *Izv. AN SSSR. Fysika Atmosfery i Okeana. (Physics of Atmosphere and Ocean)*, **19**, No. 11, 1123–1139 (in Russian).

[74] Cherny, I.V. and Etkin, V.S. (1983) High-order resonance phenomena in the

microwave emission and backscattering of ocean surface, *Dokl. Akad. Nauk SSSR*, **272**, 852–854 (in Russian).

[75] Korn, G.A. and Korn, T.M. (1968) *Mathematical Handbook.* New York: McGraw-Hill.

[76] *Radar Handbook* (1970) Ed. Scolnik, M.I., Vol. 1, New York: McGraw-Hill.

[77] Rostov, I.D. and Zhabin, I.A. (1986) Background stratification conditions for developing of double-diffusive convection in Northern Pacific, *Morskoy Gidrofizichesky Zhurnal (Marine Hydrophysical Journal)*, No. 1, 36–41 (in Russian).

[78] Kamenkovich, V.M., Koshlyakov, M.N. and Monin, A.S. (1987) *Synoptic Eddies in the Ocean.* Leningrad: Gidrometeoizdat, 510 pp. (in Russian).

[79] Legeckis, R. (1978) A survey of worldwide sea surface temperature fronts detected by environmental satellites, *J. Geophys. Res.*, **83**, No. C9, 4501–4522.

[80] Bulatov, N.V. and Lobanov, V.B. (1983) Investigations of mesoscale eddies easterly from Kuril Islands with meteorological satellites, *Issledovanie Zemli iz Kosmosa (Earth Obs. Rem. Sens)*, No. 3, 40–47 (in Russian).

[81] Ginsburg, A.I. and Fedorov, K.N. (1984) Mushroom-formed currents in ocean. *Issledovanie Zemli iz Kosmosa (Earth Obs. Rem. Sens)*, No. 3, pp. 18–26 (in Russian).

[82] Ginsburg, A.I. and Fedorov, K.N. (1984) Evolution of mushroom-like oceanic currents, *Dokl. Akadem. Nauk SSSR*, **276**, No. 2, 481–484 (in Russian).

[83] Ginsburg, A.I. and Fedorov, K.N. (1984) Some consistencies in the development of "mushroom-like" currents in the ocean revealed by analysis of space imagery, *Issledovanie Zemli iz Kosmosa (Earth Obs. Rem. Sens.)*, No. 6, pp. 3–12 (in Russian).

[84] Fu, L.L. and Holt, B. (1983) Some examples of detection of oceanic mezoscale eddies by the Seasat synthetic aperture radar, *J. Geophys. Res.*, **88**, No. C3, 1844–1852.

[85] Tai, C.K. and White, W. (1990) Eddy Variability in the Kuroshio extension as revealed by satellite altimetry: energy propagation away from the jet, *J. Phys. Oceanogr.*, **20**, 1761–1777.

[86] Hansen, D.V. and Maul, G.A. (1991) Anticyclonic current Rings in the eastern tropical Pacific. *J. Geophys. Res.*, **96**, 6965–6980.

[87] Delcroix, T., Picaut, J. and Eldin, G. (1991) Equatorial Kelvin and Rossby waves evidenced in the Pacific Ocean through Geosat sea-level and surface-current anomalies, *J. Geophys. Res.*, **96**, 3249–3262.

[88] Forbes, C., Leaman, K., Olson, D. and Brown, O. (1993) Eddy and wave dynamics in the South Atlantic as diagnosed from Geosat altimetry data, *J. Geophys. Res.*, **98**, 12297–12314.

[89] Cherny, I.V. Gus'kov, G.Ya. and Shevtsov, V.P. (1992) Microwave observations of the Kuroshio Ring, *Proceedings of IGARSS'92 Symposium*, Houston, Texas, IEEE 92CH3041-1, Vol. I, 1992, pp. 254–256.

[90] Cherny, I.V. and Shevtsov, V.P. (1995) Manifestation of synoptic eddy in the field of intrinsic microwave radiation of the ocean surface, *Earth Obs. Rem. Sens.*, **12**, 484–494 (translated from Russian).

[91]   Salomatin, A.S., Shevtsov, V.P. and Yusupov, V.I. (1985) Acoustic scattering on fine structure of hydrophysical fields, *Akust. Zh. (Acoustic Journal)*, **31**, No. 6, 768–774 (in Russian).

[92]   Shevtsov, V.P. and Salomatin, A.S. (1992) Influence of baroclinicy on the acoustic scattering properties of fine structure of hydrophysical fields in the ocean, *II Okeanologiya*, **32**, No. 4, 661–666 (in Russian).

[93]   Lobanov, V.B., Rogachev, K.A., Bulatov, I.V., *et al.* (1991) Long period evolution of warm Kuroshio eddy. *II Dokl. Akad. Nauk SSSR*, **317**, No. 4, 984–988 (in Russian).

[94]   Berestov, A.L. (1981) Some new solutions for Rossby Soliton, *Izvestiya AN SSSR, FAO (Physics of Atmosphere and Ocean)*, **17**, 82–87 (in Russian).

[95]   Berestov, A.L. (1985) Dispersion relations for Rossby Soliton, *Izvestiya AN SSSR, FAO (Physics of Atmosphere and Ocean)*, **21**, No. 3, 332–334 (in Russian).

[96]   Kizner, Z.I. (1984) Rossby Soliton with the axi-symmetrical barocline modes, *Dokl. Akadem. Nauk SSSR*, **275**, No. 6, 1495–1498 (in Russian).

[97]   Rogachev, K., Karmak, E., Miyaki, M., *et al.* (1992) Drifting buoy in the anticyclonic Oyashio eddy, *II Dokl. Russ. Acad. Sci*, **326**, No. 3, 547–550.

[98]   Gray, W.M. (1968) A global view of the origin of tropical disturbance and storm, *Mon. Wea. Rev.*, **96**, 669–700.

[99]   Ivanov, V.N., Pudov, V.D. (1977) The thermal trace of typhoon 'Tess' structure in the ocean and some parameters of energy exchange estimation during storm conditions, in *"Typhoon-75"*, Vol. 1, Leningrad: Gidrometeoizdat, pp. 66–82 (in Russian).

[100] Fedorov, K.N., Varfolomeev, A.A., Ginzburg, A.I., *et al.* (1979) Thermal reaction of the ocean on typhoon "Ella" passing. *Okeanologiya*, **19**, No. 6, 992–1001 (in Russian).

[101] Gray, W.M. (1979) Hurricanes: their formation, structure and likely role in the tropical circulation, *Meteorology over tropical oceans*, Roy. Met. Soc., **105**, 155–218.

[102] Gray, W.M. (1981) Recent advances in tropical cyclone research from radiosonde composite analysis, Dept. Atm. Sci., Colorado State Univ., 407 pp.

[103] Martin, J.D. and Gray, W.M. (1993) Tropical cyclone observation and forecasting with and without aircraft reconnaissance, *Wea. Forecasting*, **8**, 519–532.

[104] Willoughby, H.E. (1995) Mature structure and evolution, *Global Perspectives on Tropical Cyclones*, Elsberry, R.L. (Ed.). World Meteorological Organization, Report No. TCP-38, Geneva, 62 pp.

[105] Zehr, R.M. (1992) Tropical cyclogenesis in the western North Pacific, *NOAA Technical Report NESDIS 61*, U.S. Department of Commerce, Washington, DC 20233, 181 pp.

[106] Cherny, I.V. (1993) Microwave studies of typhoon *Warren* generation, *Proceedings of IGARSS'93 Symposium*, Tokyo, IEEE 93CH3294-6, Vol. 4, pp. 1913–1915.

[107] Khain, A.P. and Sutyrin, G.G. (1983) Tropical cyclones and its interactions with the ocean. Leningrad: Gidrometeoizdat, 272 pp. (in Russian).

[108] Pereskokov, A.I. and Fedorov, K.N. (1985) Double-diffusive convection in the

ocean as a climate forming factor, *DAN SSSR*, **285**, No. 1, 229–232 (in Russian).

[109]   Fedorov, K.N. (1984) Stratification conditions and salt fingers convection in the ocean, *DAN*, **275**, No. 3, 749–753 (in Russian).

[110]   Fedorov, K.N. (1986) Layers thickness and interchange coefficients during interleaving convection in the ocean, *DAN*, **287**, No. 5, 1230–1233 (in Russian).

[111]   Schmitt, R.W. (1981) Form of the temperature–salinity relationship in the central water evidence for double-diffusive mixing, *J. Pgys. Oceanogr.*, **11**, No. 7, 1015.

[112]   Esersky, A.B., Korotin, P.I. and Rabinovich, M.I. (1985) Chaotic auto-modulation of two-measured structures on liquid surface by parametric stimulation, *Pis'ma v ZHETF (Letters in Journal of Experimental and Theoretical Physics)*, **41**, 129–131 (in Russian).

[113]   Esersky, A.B., Rabinovich, M.I., Reutov, V.P. and Starobinets, I.M. (1986) Spatial–temporal chaos of capillary ripples by parametric stimulation of the liquids, *ZHETF (Journal of Experimental and Theoretical Physics)*, **91**, No. 6(12), 2070–2083 (in Russian).

[114]   Sitnikov, I.G. (1987) Forecast of tropical cyclones: modern state and perspectives, *Meteorology and Hydrology*, No. 2, 115–121 (in Russian).

[115]   Sitnikov, I.G. (1991) Numerical forecast of tropical cyclones movement: successes for five recent years, *Fifth International Symposium on Tropical Meteorology*, May–June, Moscow: Obninsk, p. 13 (in Russian).

[116]   Gray, W.M., Landsea, C.W., Mielke, Jr., P.W. and Berry, K.J. (1992) Predicting Atlantic seasonal hurricane activity 6–11 months in advance, *Wea. Forecasting*, **7**, 440–455.

[117]   Bender, M.A., Ross, R.J., Tuleya, R.E. and Kurihara, Y. (1993) Improvements in tropical cyclone track and intensity forecasts using the GFDL initialization system, *Mon. Wea. Rev.*, **121**, 2046–2061.

[118]   Fiorino, M., Goerss, J.S., Jensen, J.J. and Harrison, Jr., E.J. (1993) An evaluation of the real-time tropical cyclone forecast skill of the Navy operations global atmospheric prediction system in the Western North Pacific, *Wea. Forecasting*, **8**, 3–24.

[119]   Lander, M. (1994) An exploratory analysis of the relationship between tropical storm formation in the Western North Pacific and ENSO, *Mon. Wea. Rev.*, **122**, 636–651.

[120]   Landsea, C.W., Gray, C.W., Mielke, Jr., P.W. and Berry, K.J. (1994) Seasonal forecasting of Atlantic hurricane activity, *Weather*, **49**, 273–284.

[121]   Neumann, C.J., Jarvinen, B.R., McAdie, C.J. and Elms, J.D. (1993) Tropical cyclones of the North Atlantic Ocean, 1871–1992, Prepared by the National Climatic Data Center, Asheville, NC, in cooperation with the National Hurricane Center, Coral Gables, Fl, 193 pp.

[122]   Tuleya, R.E. (1994) Tropical storm development and decay: Sensitivity to surface boundary conditions, *Mon. Wea. Rev.*, **122**, 291–304.

[123]   Pudov, V.D., Svirkunov, P.N. and Teslenko, V.P. (1991) General results of Soviet complex expedition "Typhoon-90", *Fifth International Symposium on Tropical Meteorology*, May–June, Moscow: Obninsk, p. 7 (in Russian).

[124]   Elsberry, R.L. (1991) Short report of field experiment on studying the tropical

cyclones (TCM-90), *Fifth International Symposium on Tropical Meteorology*, May–June, Moscow: Obninsk, p. 5 (in Russian).

[125] Ivanov, V.N., Pudov, V.D. and Fedorov, K.N. (1980) On the generation of oceanic cyclonic eddy under typhoon influence, *DAN SSSR*, **253**, No. 6, 1454–145 (in Russian).

[126] Sutyrin, G.G. (1981) On the development of cyclonic eddy in the stratified ocean under influence of tropical cyclone, *Okeanologiya*, **21**, No. 1, 12–18 (in Russian).

[127] Pudov, V.D. (1987) Winter transformation of thermic regime of upper ocean layer in the region of typhoons origination and evolution, *Tropical Meteorology, Proceedings of Third International Symposium*, Leningrad: Gidrometeoizdat, pp. 374–379 (in Russian).

[128] Egrenich, E.A. (1982) Influence of ocean surface temperature nonhomogeneities on the movement and evolution of tropical cyclones, *Tropical Meteorology, Proceedings of Third International Symposium*, Leningrad: Gidrometeoizdat, pp. 384–390 (in Russian).

[129] Pokhil, A.E. and Chernyavsky, E.B. (1986) On the reaction of the ocean on tropical cyclone origination and movement, *Meteorology and Hydrology*, No. 7, 55–61 (in Russian).

[130] Sitnikov, I.G., Zlenko, V.A., Pokhil, A.E. and Yurko, T.A. (1989) Preliminary results of operative testing of a numerical technique for predicting the tropical cyclones movement and influence of the ocean on their tracks, *Tropical Meteorology, Proceedings of Fourth International Symposium*, Leningrad: Gidrometeoizdat, pp. 186–198 (in Russian).

[131] Tropical Cyclones. (Eds Voloschuk, V.M. and Nerushev, A.F.), Leningrad: Gidrometeoizdat, 1989, 54 pp. (in Russian).

[132] Pavlov, N.I. (1985) Peculiarities of typhoon movement onto Soviet Far East, *Tropical Meteorology, Proceedings of Second International Symposium*, Leningrad: Gidrometeoizdat, pp. 80–86 (in Russian).

[133] Cherny, I.V. (1993) Microwave diagnostics of tropical cyclone anomalous trajectory. *Proceedings of IGARSS'93 Symposium*, Tokyo, IEEE 93CH3294-6, Vol. 4, pp. 2151–2154.

[134] Dobrovol'sky, A.D. and Zalogin, B.S. (1992) *Regional Oceanology*. The Moscow State University Edition, Moscow, 224 pp. (in Russian).

[135] Wentz, F.J. (1991) User's Manual SSM/I Antenna Temperature Tapes Revision 1, *RSS Technical Report 120191*, Dec. 1, Remote Sensing Systems, Santa Rosa, CA.

[136] Wentz, F.J. (1993) User's Manual SSM/I Antenna Temperature Tapes Revision 2, *RSS Technical Report 120193*, Dec. 1, Remote Sensing Systems, Santa Rosa, CA.

[137] Wentz, F.J. (1990) SBIR phase II report: West coast storm forecasting with SSM/I, *RSS Technical Report 033190*, 378 pp., Mar., Remote Sensing Systems, Santa Rosa, CA.

[138] Perry, A.H. and Walker, J.M . (1977) The ocean–atmosphere system. London and New York: Longman.

[139] Groen, P. (1967) *The Waters of the Sea*, Van Nostrand, 328 pp.

[140] The structure and variability of large-scale oceanologic processes and fields in the

Norwegian energy active zone (1989) Eds By Nikolaev, Yu.V. and Alexeev, G.V. Leningrad: Gidrometeoizdat, 128 pp. (in Russian).

[141]   Monin, A.S. (1978) *Oceanologiya. Hydrophysics of Ocean*, Vol. 1, Moscow: Nauka, 456 pp. (in Russian).

[142]   Grankov, A.G. and Usov, P.P. (1994) The relation of monthly water–air temperature difference and the characteristics of thermal emission of ocean and troposphere in microwave and IR-band, *Meteorology and Hydrology*, No. 6, 79–89 (in Russian).

[143]   Grankov, A.G. and Resnyanskii, Yu.D. (1997) Modeling the response of the internal irradiation of the ocean–atmosphere system to the perturbation of a thermal equilibrium at the interface, *Meteorology and Hydrology*, No. 11, 78–89 (in Russian).

[144]   Laevastu, T. and Hela, I. (1970) Fisheries oceanography in *Fishing News* (Book) LTD., London.

Conclusions

Nowadays, it is beyond question that the study of the ocean and the atmosphere as a coupled system can no longer be carried out without remote sensing means, among which the passive microwave means and techniques are very effective due to their high sensitivity to the variability of hydrometeorological parameters. The main goal of this monograph has been to describe in more detail the principles of microwave diagnostics of the processes and phenomena on the ocean surface and, in addition, to demonstrate the possibilities of microwave technique for studying the ocean active layer processes, including those producing the decisive influence on the atmosphere state. From our point of view, the material presented reflects the whole history of the development of this branch of radiohydrophysics.

Microwave characteristics of the ocean surface are defined by three factors: dielectric properties of water skin depth; surface wave geometry; and dispersive structures, appearing after wind wave breaking. The contributions of geometric and volume non-homogeneities in the microwave emission of the ocean surface are different. The variations of microwave emission caused by geometric factors depend on spatial structure and scale of roughness. The small perturbations method, based on the resonant theory, remains quite acceptable for the explanation of the brightness temperature contrast of ocean surface at millimeter- and centimeter-wave frequencies under low wind speed. At the same time, the statistical averaging made in the frame of a two-scale model using the geometric optics approach results in more strict issues mainly under the grazing viewing angles. The influence of dispersive formation or other volume non-homogeneities on the spectrum of the microwave emission of the ocean surface could be taken into account by using a single macroscopic approach based on the models of layer-inhomogeneous media with distributed parameters. The developed multiparametric radiophysical model of the ocean, considering the additional contribution of different factors and its spatial variability in the wind field, introduces a new approach for the interpretation of microwave remote sensing data.

The results of field experiments presented in this monograph to some extent violate the traditional opinion that microwave radiometry and, in particular, millimeter-wave frequencies are not useful for the remote sensing of the oceans, mainly because of the heavy attenuation of electromagnetic waves in the atmosphere. In fact, the highly selective character of the microwave emission of the ocean surface and therewith a large contrast in brightness temperature variations up to 20–30 K caused by a number of oceanic processes were somewhat unexpected. Using the said electrodynamic sign we combined the different processes in the ocean–atmosphere system from small-scale 50–100 m ('relic rain') to mesoscale 10–60 km (frontal zone) and synoptic 200–650 km (oceanic synoptic eddies, oceanic fronts, and synoptic vortexes in the atmosphere), and tried to establish in a qualitative manner their relation to the development of instability of ocean thermohaline fine structure. Just such an instability could serve as a litmus test, providing the remote diagnostics of deep ocean processes, of critical situations in atmosphere, and of natural disasters.

In general, the experimental results of field investigations discussed demonstrate the possibilities of the passive microwave technique for the remote sensing of the oceans and confirm the original approach for microwave diagnostics of the deep ocean processes, based on an amplification mechanism concept. Nevertheless, the interdisciplinary approach and further researches are needed to construct a full physical model of observed phenomena, both electrodynamic and hydrophysical.

In this connection, the development of advanced multichannel microwave imaging radiometers to be deployed on research aircraft and the 'Meteor-3M' satellite, providing multispectral microwave measurements available by no other means, offers a fertile field for further research.

Index